Complex Terrain Mapping

Lambert Rivard • Carla Hehner-Rivard

Complex Terrain Mapping

Integrated Use of Stereo Air Photos and Satellite Images

Lambert Rivard
Carla Hehner-Rivard
Saint-Lambert, QC
Canada

ISBN 978-3-319-02449-3 ISBN 978-3-319-02450-9 (eBook)
DOI 10.1007/978-3-319-02450-9
Springer Cham Heidelberg New York Dordrecht London

Library of Congress Control Number: 2014940976

Printed on acid-free paper

Springer is part of Springer Science+Business Media (www.springer.com)

Preface

This monograph documents a concept of complex terrains with the use of 57 sites consisting of combined interpreted stereo air photos and related Landsat images of each site. These are complemented by 21 field photos of slope failures.

Nature of Complex Terrain

The term *complex terrains* is adapted from D. H. Huntley (2008) *Landslide Geohazard Mapping in Complex Terrains*, Geological Survey of Canada Open File 5747. The term is paraphrased here to define regions of interaction of complex tectonic and geomorphic history, a range of surficial deposits and bedrock, as well as high relief and steep slopes.

Conventional Stereo air photos

Black-and-white, color, and false-color stereo air photographs continue to be routinely taken by many governments and remain the preferred medium for producing maps that depict "the distribution, extent and location of bedrock, earth materials, landforms and geomorphic processes" (Huntley). The air photos in this work range in scale from 1: 10,000 to 1: 125,000.

High-Resolution Stereo Satellite Images

These images, state of the art for many Earth science applications, are equally applicable to mapping of complex terrains, but the use of conventional air photos remains competitive. The paper by J. E. Nichol et al (2006) *Application of high resolution stereo satellite images to detailed landslide hazard assessment*—Geomorphology vol. 76, pp 68–75, included a cost–benefit analysis comparing stereo air photo interpretation with stereo satellite interpretation. Although it suggested that stereo satellite imagery is more cost-effective for detailed landslide hazard assessment over large areas, the analysis showed (Table 2) that the total job cost (cost of photos/images plus cost of interpretation and mapping) for a landslide survey of a 1,000 km^2 area based on 1:10,000 stereo air photos was US$30,200 and US$48,720 for 0.87 resolution IKONOS image use. Material cost of air photos was US$11,000 and US$46, 000 for IKONOS images.

Saint-Lambert, QC, Canada Lambert Rivard

Contents

Introduction

Mapping Method

The mapping of complex terrains, following background research and appreciation of regional environment on Landsat imagery is based on the viewing of stereo air photos to delineate boundaries between surface areas of geounits of different types **detected** on the basis of **spectral and spatial elements**, and **identified** and **interpreted** by associational elements according to a genetic classification.

Photo Geounit Detection Characteristics

Spectral elements are tone and texture:

- **Tone** is the relative brightness of a unit expressed in levels of grey in black and white photos
- **Texture** is the relative roughness/smoothness of a unit surface produced by the frequency of tonal change in the unit
- **Spatial** elements are relief, shape, drainage, and land use or land cover
- **Relief** is expressed by stereo viewing in air photographs, and by sun elevation, azimuth and shadowing in monoscopic satellite images
- **Shape** is the characteristic shape of some geounits
- **Drainage** is the density and arrangement of stream channels within a unit
- **Land use and land cover** generally occur distinctively in a unit

Identification and Interpretation of Photo Geounits

Geounits are identified by **Associational Patterns,** of a number of spectral and spatial elements that correlate in an air photo to characterize a homogeneous geounit that can be deductively **interpreted** according to a **genetic classification** of accepted geoscience usage. Geological and geomorphological mapping requires

L. Rivard and C. Hehner-Rivard, *Complex Terrain Mapping*,
DOI 10.1007/978-3-319-02450-9_1, © Springer International Publishing Switzerland 2014

the use of a set of descriptor codes to apply to mapped geounits. The complete set of descriptors constitutes a genetic classification system. When geounits are compiled as a photogeologic map these codes become the map legend.

The stereo pair of Figure B7 illustrates the application of these elements:

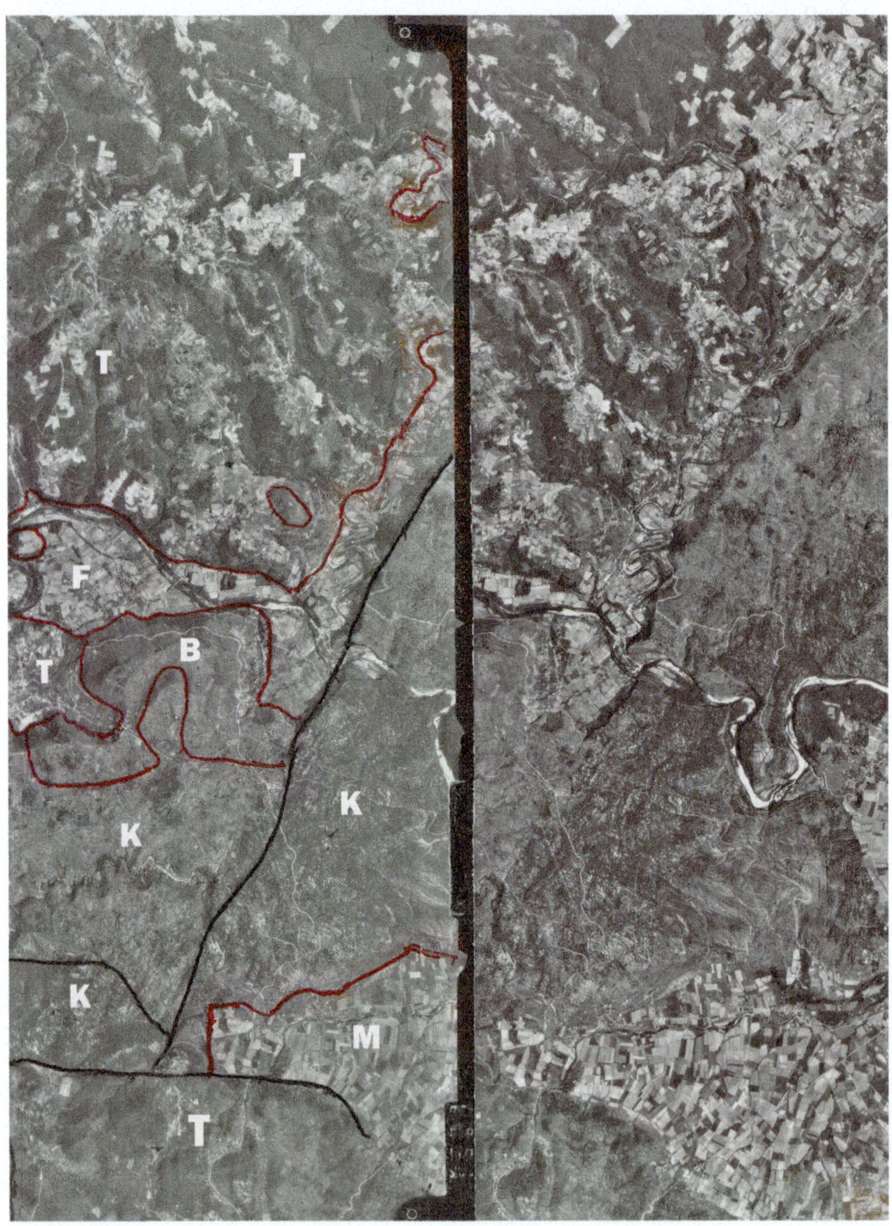

Tone is bright in the agricultural areas of the **F** river valley terraces and **M** marl lowland in the southeast corner
Texture is rough on the limestones at **K** and smooth at **M**
Relief is in the stereo contrast between the **T** Triassic upland and the **K** limestones
Shape is the **F** typical broad terraces bordering the river floodplain
Drainage is the stream system dissecting the interbedded **T** Triassic sediments
Land use and land cover—Agriculture on the interfluves and forest cover on the slopes of unit **T**
Association Pattern—Unit **K** is an association of tone, texture relief and land cover; Unit **T** is an association of tones, relief drainage, land use and land cover; Unit **P** is an association of relief and land cover; Unit **F** is an association of tone, relief, shape, and land use

Given the aptitude of the person performing an interpretation the expression of a geounit within a photo and its mappability is ultimately conditioned by three fundamental factors:

- Sufficient distinction in lithologic composition or structure occurring in a terrain based on the detection and identification elements
- The nature of the denudational processes that act or have acted on it
- The biophysical environment in which it is presently occurring

S.A. Drury (1987) wrote in his text *Image Interpretation in Geology*, p145, *"The problem with geological patterns is that as well as being complex in their own right, they are also part of a complex whole."*

Geologic Time Nomenclature

Table 1 is a reference scale for the time terms of geologic processes and units used in the Examples. The **Hercynian Orogeny** was throughout the Paleozoic Era; the **Alpine Orogeny** was throughout the Mesozoic and Cenozoic Eras.

Terrain Type Categories

A review of the author's archives produced a scope of five major categories of morphogenetic terrain types and processes that meet the complex terrain criteria.

- Bedrock terrains, 3 types, 11 examples
- Glacial and alpine-glacial terrains, 13 types, 15 examples
- Periglacial terrains, 3 types, 6 examples
- Nonglacial surficial terrains, 4 types, 4 examples
- Landslide terrains, 7 types, 21 examples
- Field photos of slope failures, 21 photos of 13 sites

Table 1 Geological time scale

EON	ERA	PERIOD	Ma	EPOCH	
PHANEROZOIC	**Cn** CENOZOIC	**Q** Quaternary	2	**R** Holocene	10 ka
				Würm	–80–10 ka
				Pl Pleistocene	130 ka
		T Tertiary	65	**Po** Pliocene	5 Ma
				Mc Miocene	25 Ma
				Og Oligocene	34 Ma
				E Eocene	55 Ma
				Pe Paleocene	65 Ma
	M MESOZOIC	**K** Cretaceous	140	**Ku** Upper	95 Ma
				Kl Lower	140 Ma
		J Jurassic	200		
		Tr Triassic	240		
	PZ PALEOZOIC	**Pm** Permian	300		
		Cb Carboniferous	350	**Cbu** Upper	
				Cbl Lower	
		D Devonian	410		
		S Silurian	440		
		O Ordovician	510		
		C Cambrian	545		
Pr PROTEROZOIC			2500		
Ar ARCHEAN			4000		

Air Photo Layout Format

The air photos of all Figures are arranged as pairs. A 16 by 22 cm portion of a conventional 23 by 23 cm air photo area is reproduced first to conform to the 26 by 19 cm publication format.

The second photo is an 8 cm stereo/strip using the mate of the first photo.

Regional Setting Satellite Images

A portion of a 15 or 30 m resolution Landsat scene selected from USGS/EROS Global Visualization Viewer archives is interpreted to provide the regional structural and physiographic setting of each framed example.

A zoom of detail of a particular Example air photo is readily visualized by a searcher using Google Earth™ satellite data.

Complex Terrain Categories

Bedrock Terrains

Eleven Examples of three rock classes include three of crystalline terrain; one of volcanic structure, and seven of sedimentary successions. Five of the latter are in southern France where complex associations of high density land uses and morphological units in developed areas are illustrated.

L. Rivard and C. Hehner-Rivard, *Complex Terrain Mapping*,
DOI 10.1007/978-3-319-02450-9_2,

Figure B1: Mecatina

Location 50°49′N, 59°58′W

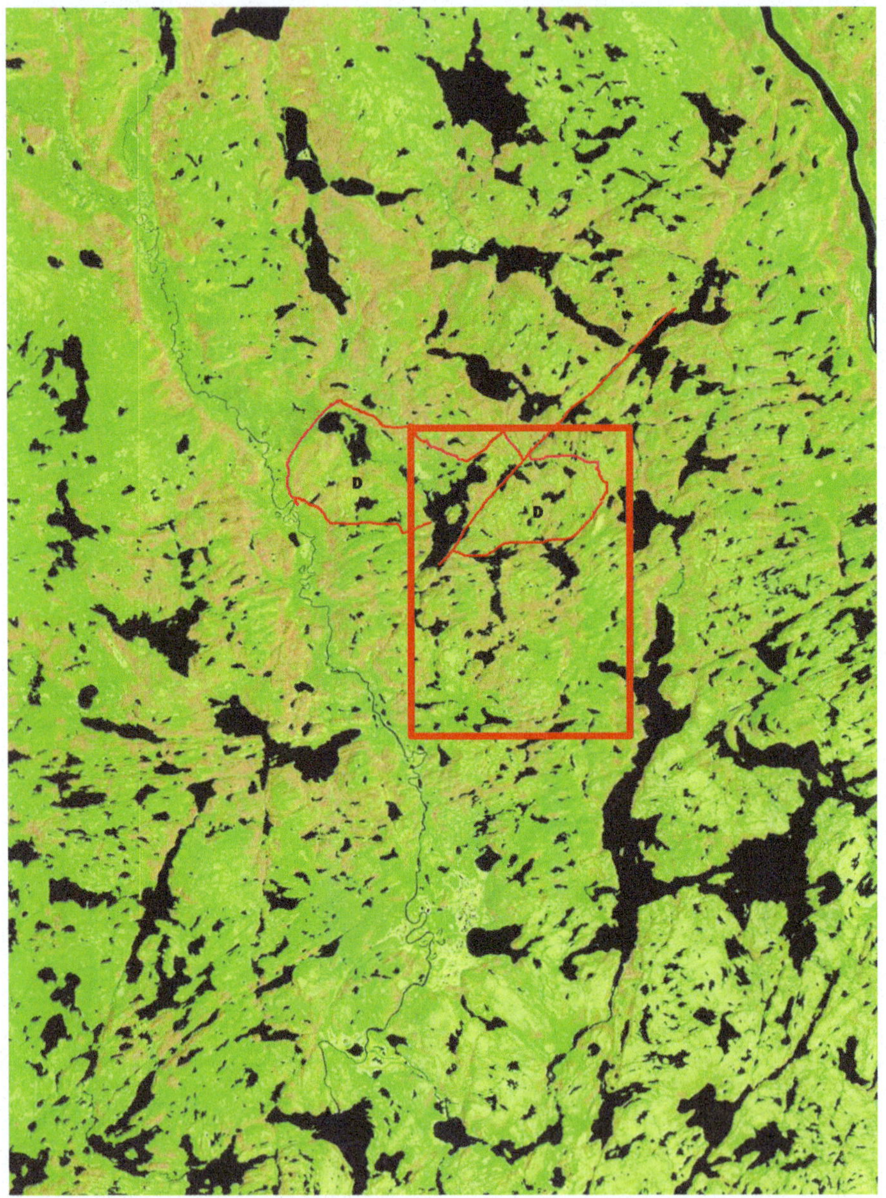

Landsat Image TM 4/5, 27 October 2009

Regional Environment

The 1,750 km^2 image at 1:220,000 scale is at the extreme east limit of the Province of Quebec, near the junction of the Gulf of St. Lawrence and the Strait of Belle Isle.

Structure

The area is in the geologically complex Grenville Province crystalline terrain of the southeast Canadian Shield. Two structural features related to the air photo area have been drawn on the Landsat image. A 20 km long northeast-trending lineament is interpreted as a strike-slip fault offsetting a **granite gneiss dome** outlined on either side of the fault-controlled lake.

Physiography

The area is a lake-studded plateau of rough and irregular relief of Precambrian peneplanation and Pleistocene glaciations, ranging from 360 to 550 m in elevation. The entire area is forested in relatively open stands of boreal coniferous black spruce.

Rm
Rm
4
14
4
4
14
4
Rm
Rm
Rm
6
Rm
Rm
Rf
4
4
4
Rf
4
Rf
Rf
10
Rf
Rf
Rf
Rm
Rm
Rf
Rm
4
4
Rm
14
Rm
Rm
Rm
Rm
6
Rm
14
Rf
Rm

B1 Airphoto

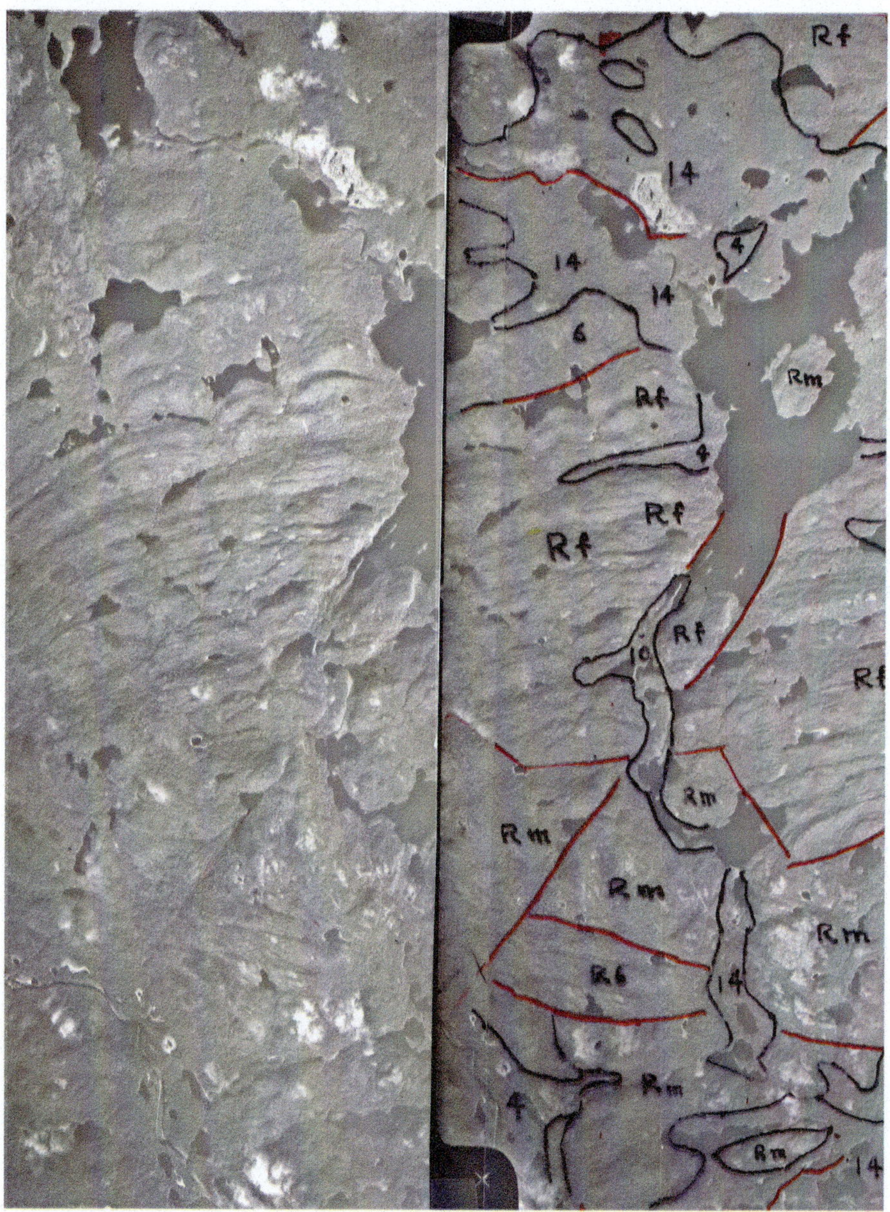

B1 Stereopair

Photo Interpretation

The framed area of this 1:60,000 scale air photo is 160 km^2. Twenty-four lithomorphic geounits have been delineated and coded in four genetic categories and seven lithomorphic types. Some undivided lineaments are drawn in red.

Bedrock Units

Rf—These extensive units of Proterozoic gneiss have the typical morphology of parallel linear ridges reflecting the foliated banding of the rock. The relief of individual ridges ranges from 20 to 40 m.

Rm—Twelve occurrences of massive granite are delineated. The three largest occupy the north quarter of the photo, while the rest are lineament-bound in the southwest.

Glacial Deposits

4—Four units of ground moraine occur in limited areas of narrow depressions in bedrock.

6—Three units of veneer moraine occur as areas adjacent to units of outwash.

Glaciofluvial Deposits

10—An esker-outwash complex occupies a 3 km long 300 m wide depression threaded between two lakes.

Proglacial Deposits

14—Three outwash plains are delineated. The largest, crossed by an esker, is 9 km^2 adjacent to the lake in the northwest.

Photo Source

Courtesy of National Air Photo Library, A20256, 21–22

References

Davidson, A. (1996). Geology of Grenville Province, Geological Survey of Canada, Open File 3346, scale 1:2,000,000.

Les parcs québécois (undated), No 7, Les régions naturelles Région B 16, p 87.

Figure B2: Ellesmere

Location 80°25′N, 79°25′W

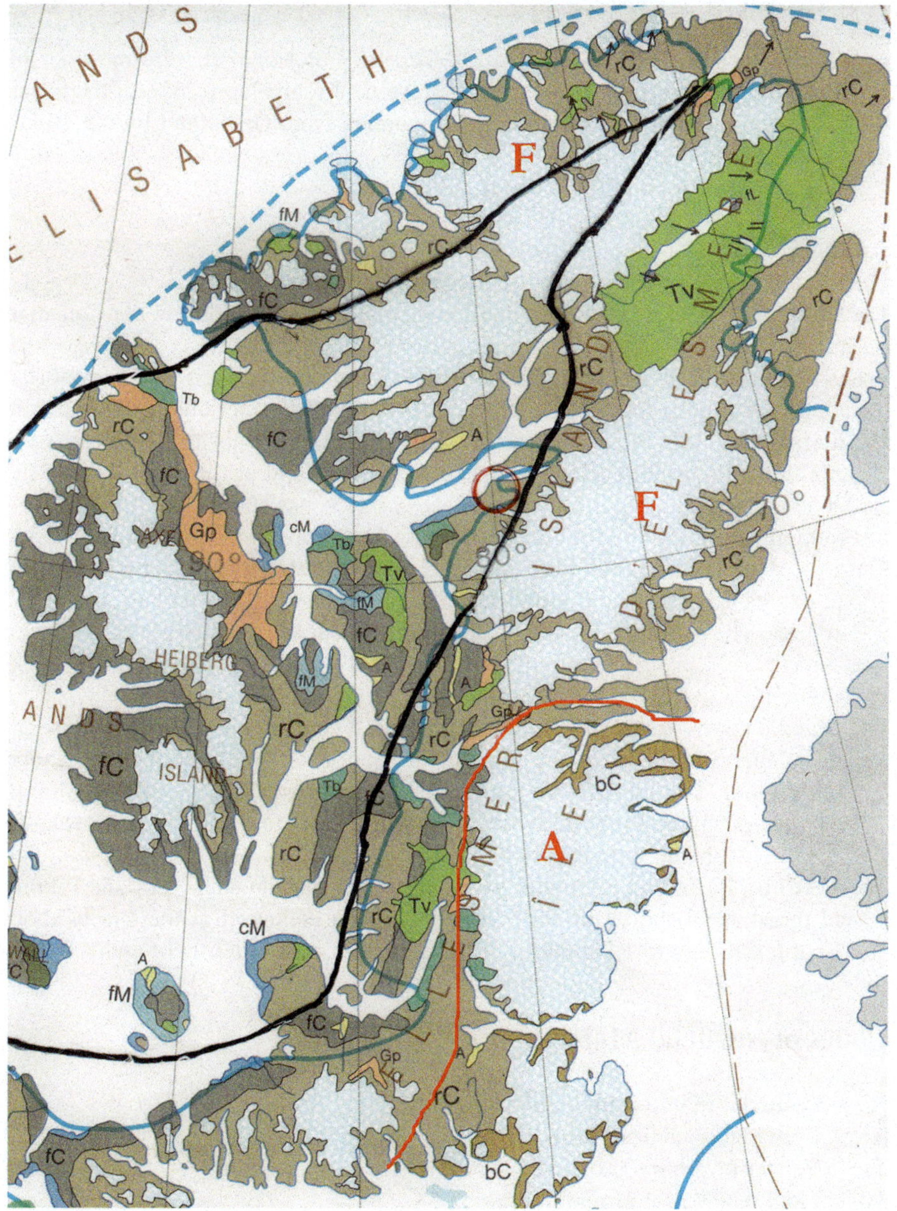

Regional Environment

No usable Landsat image could be found in the USGS/EROS Global Visualization Viewer Archive (Google Earth™ images are not used as a source). A map of Ellesmere Island is substituted. The map is taken from Surficial Materials of Canada, 1:5,000,000, GSC Map 1880A, 1995.

With an area of 196,235 km^2 in the Territory of Nunavut, Ellesmere is the northernmost of the islands of the Canadian Arctic Archipelago. 25 % of the island is covered by ice caps and glaciers. It is separated from Greenland by the 40 km wide Nares Strait.

The Example area is 150 km south of Figure E4.

Structure

The island consists of three major structural units.

The segment marked **A** on the southeast is the most northern point of the Canadian Shield.

The areas marked **F** are parts of the Franklinian Mobile Belt of Precambrian sedimentary rocks folded and thrusted during the Ellesmerian Orogeny in late Silurian. The belt is a sinuous arc zone traced from north Greenland over 2,000 km through Ellesmere Island to Melville Island at the southwest of the Arctic archipelago.

The third unit, bounded by the black line, is the Sverdrup Basin. Originally a simple crustal depression in the Franklinian rocks, it was filled with Carboniferous and Mesozoic clastic strata deformed in the early Tertiary by the Eurekan (Innuitian) Orogeny. The red circled site of the example is near the east margin of the basin.

Physiography

The island's coasts are deeply incised by fjords. The example site is on the south side of Iberville Fjord, a side fjord of the major 400 km long Nansen Sound Greely Fjord. The Franklinian terrains are mountainous, rising to over 2,500 m and covered by ice caps and outlet glaciers. These are the highest peaks in Canada outside the west coast mountains.

The Sverdrup Basin is a ridge and valley dissected plateau shaped by the folding and thrusting of the basin sediments. The basin elevation at the site is about 750 m but it rises to 1,700 and 2,500 m at its ice capped northern end.

Codes of Surficial Materials on the Map

rC Colluvial rubble from carbonate rocks
fC Colluvial fines from weak shales
Tv Veneer of ground moraine
Tb Thick blanket of ground moraine
cM Coarse-grained glaciomarine deposits
fM Fine-grained glaciomarine deposits
Gp Glaciofluvial outwash sands/gravels

R1
R2
18
18
R2
R1
R1
18
R1
R1
R2
R1
R1
18
18
R1
18
R1
18
R1
R1
18
R1
18
R2
18
R1
R2
18
18
R1
R1
R1
18
R1
18
R2
R1
R1
R1
18
R1

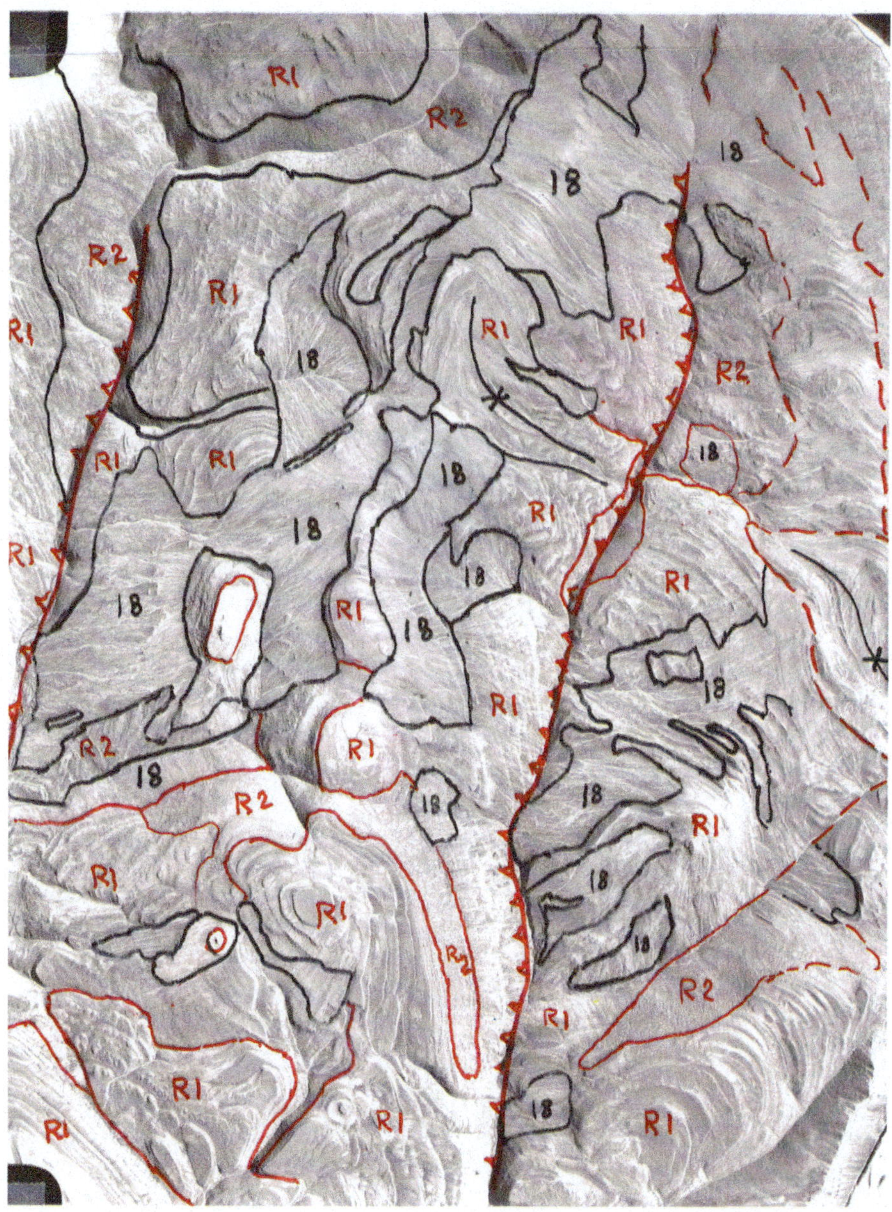

B2 Airphoto

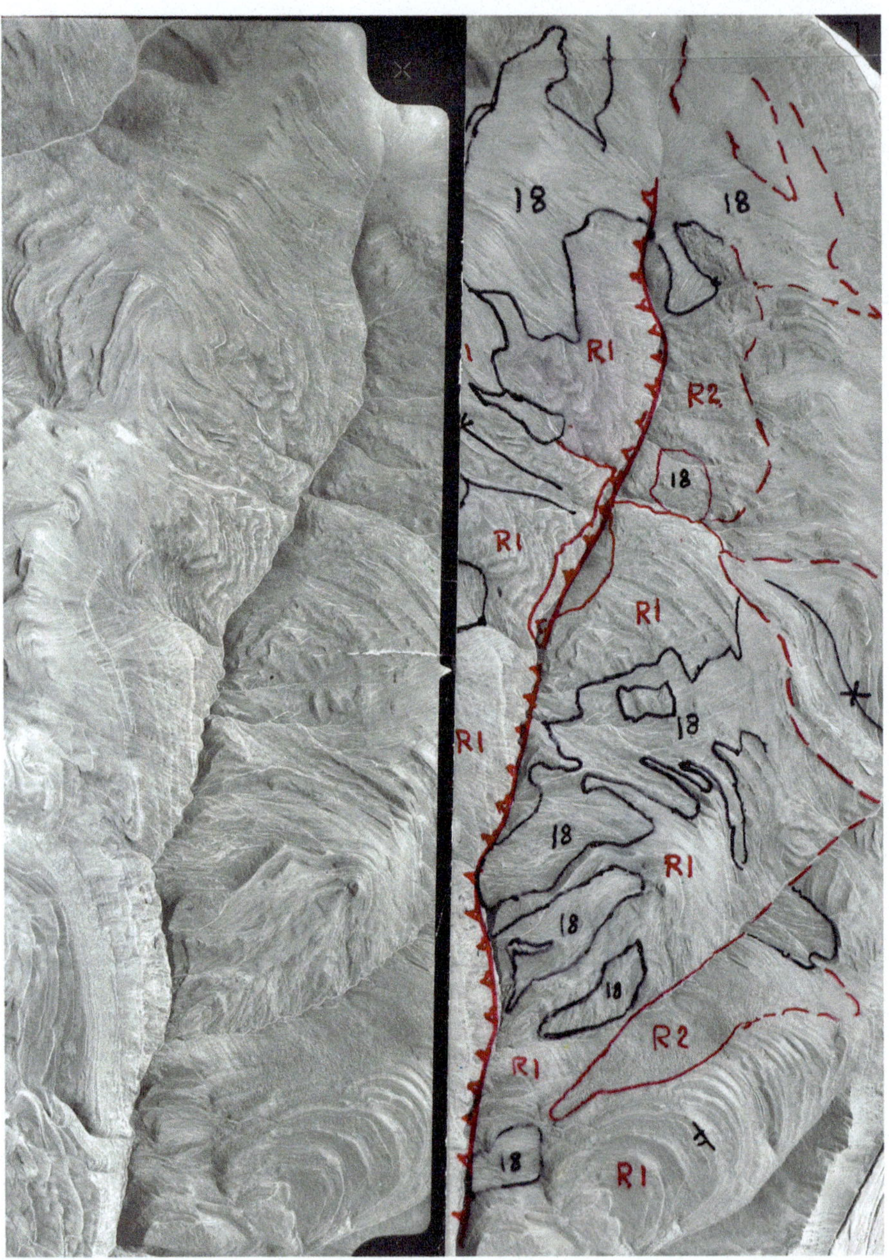

B2 Stereopair

Photo Interpretation

The three units interpreted in the 145 km^2 area of this 1:60,000 scale air photo are delineated in a complex pattern of 35 occurrences. The white area at photo bottom is a glacier.

There is no evidence of glacial erosion or deposition, the blue line on the map shows the site to be just beyond the limit of extent of the last (Wisconsinan) ice cover.

The well-exposed and locally thrust traced Sverdrup Basin sedimentary rocks in the polar desert environment are divided into two units thinly covered by unresolved colluvial rubble.

R1—The characteristic resistant-weak stepped form of the little-disturbed interbedded strata of this unit is easily mapped.

R2—These units of limited occurrence are of massive and rougher appearance in contrast with the bedded R1 units.

18—Periglacial gelifluction deposits cover fairly extensive slopes in the north, west, and east parts of the photo.

Photo Source

Courtesy of National Air Photo Library, A16693, 147–148

References

Hodgson, D. A. (1989). Quaternary geology of the Queen Elizabeth Islands, Chapter 6. In R. J. Fulton (Ed) Quaternary geology of Canada and Greenland. *Geological Survey of Canada, Geology of Canada, no. 1*, 445–447.

Wheeler, J. O., et al. (1996). *Geological map of Canada.* Geological Survey of Canada Map 1860A scale 1:5,000,000.

Figure B3: Diyagama

Location 06°49′27″, 80°48′23″E

Landsat Image 7 ETM, 14 March 2001

Regional Environment

The 1:550,000 scale subscene covers 3,465 km^2 of south central Sri Lanka's Highland Complex in which the black inset frame of the Example site is located.

Structure

The Highland Complex is part of a geologically old island consisting primarily of gneisses, metamorphosed sediments, and granite rocks of Precambrian age. It includes a group of rugged structurally controlled mountain ranges and ridges ranging from 1,300 to 2,500 m in heights.

Four geomorphic units delineated on the image define the Complex's main structural features.

A units are dominantly foliated northwest striking metasedimentay strata.

A1 unit are similar but north striking suites.

M units are massive, broadly folded and block faulted metamorphic granitoids and thick bedded quartzites.

G units are granites.

Physiography

Deep physico-chemical weathering is active throughout the region.

The terrain of the **M** unit consists of faulted hills with <90 m relief, with rounded or flat-bottomed valleys.

The dominant **A** units are a succession of parallel ridges and valleys whose orientations are controlled by the strike of the folded strata. The relief of valleys to crests ranges from 90 to 600 m.

The bit of Unit **G** terrain cut by the image frame is mountainous with hills and ridges ranging from 1,500 to 2,200 m in height.

The dark green areas on the image are dense 25–30 m high montane rain forests with an annual rainfall >1,800 mm. The region has a northeast monsoon season from December to February.

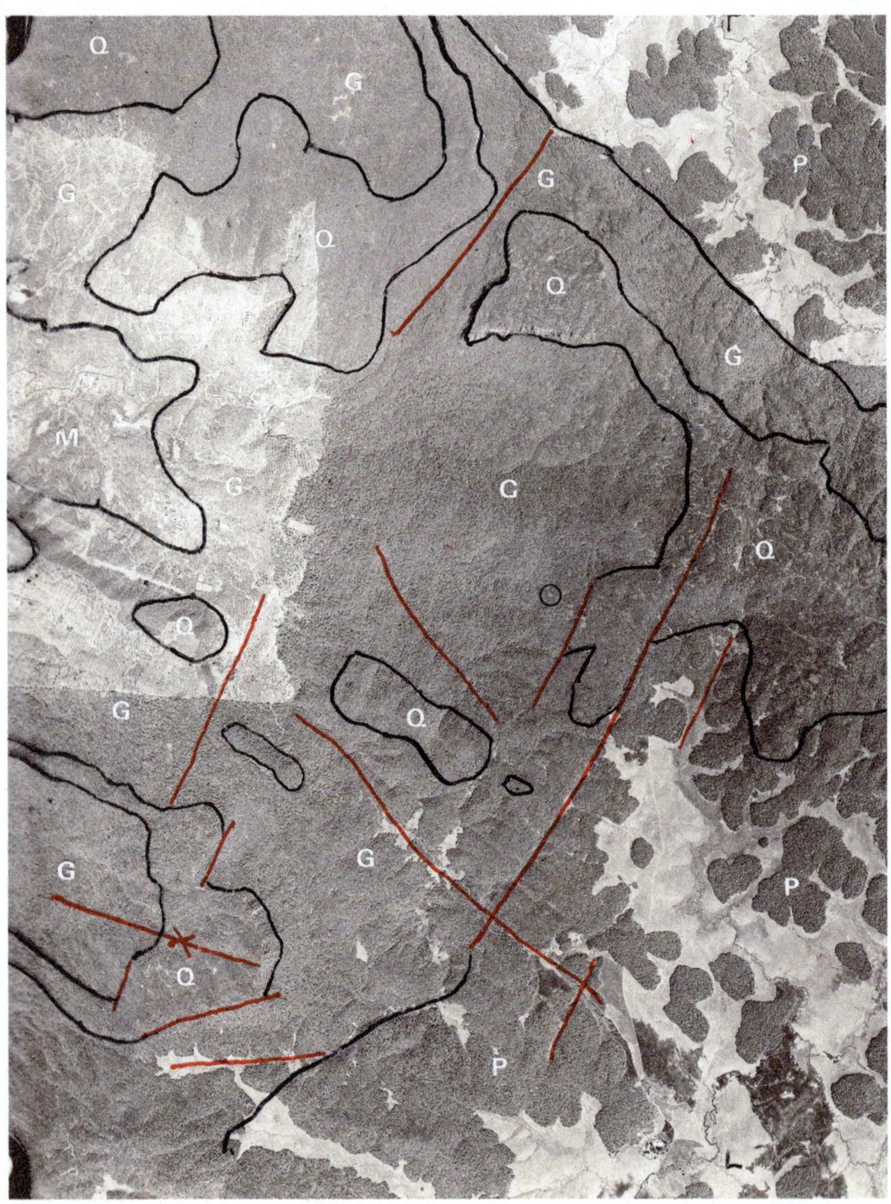

B3 Airphoto

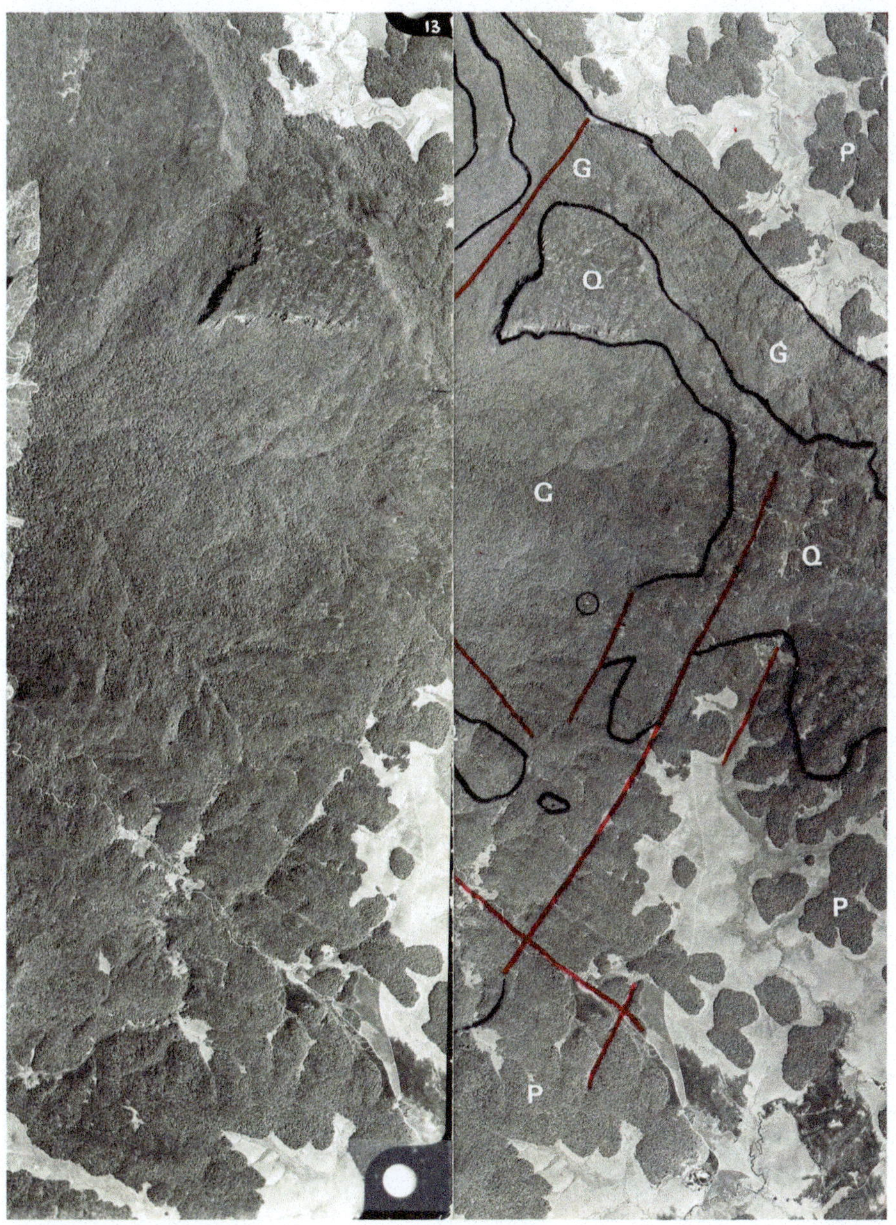

B3 Stereopair

Photo Interpretation

The area of the framed 1:40,000 scale air photo is 55 km^2.

The interpretation of the Landsat image shows the air photo site to be in the **M** unit of broadly folded granitoids of the Highland Complex. Interpretation of the photo shows four rock types and structures that make up the regional unit.

G is granite gneiss, the central part of which is the uplifted end of a 6 km broad syncline plunging gently to northwest. It rises 450 m from 1,650 m at the valley forming **M** metasediments to 2,110 m at its probably fault-bounded contact with the **P** plateau. It is dissected by streams flowing down its forested inner dipping beds into the low ground of the cleared tea estate.

Q is thick resistant flat-lying quartzite that caps the granites at 2,200 m elevation.

P at 2,150 m is an area of senile topography which shows evidence of peneplanation. The terrain consists of low relief forested rounded hills and grass-covered valleys.

A set of northeast-striking lineaments crosses the center of the photo.

Photo Source

Personal archive

References

Erb, D. K. (1970). Landforms and drainage of Ceylon. *Ceylon Geographer, 20*, 18–25.

Parkinson, R. N., et al. (1962). A report on the resources of the Mahaweli Ganga Basin Ceylon. *Canada-Ceylon Colombo Plan Project.*

Figure B4: Cannes

Location 43°43′N, 06°54′E

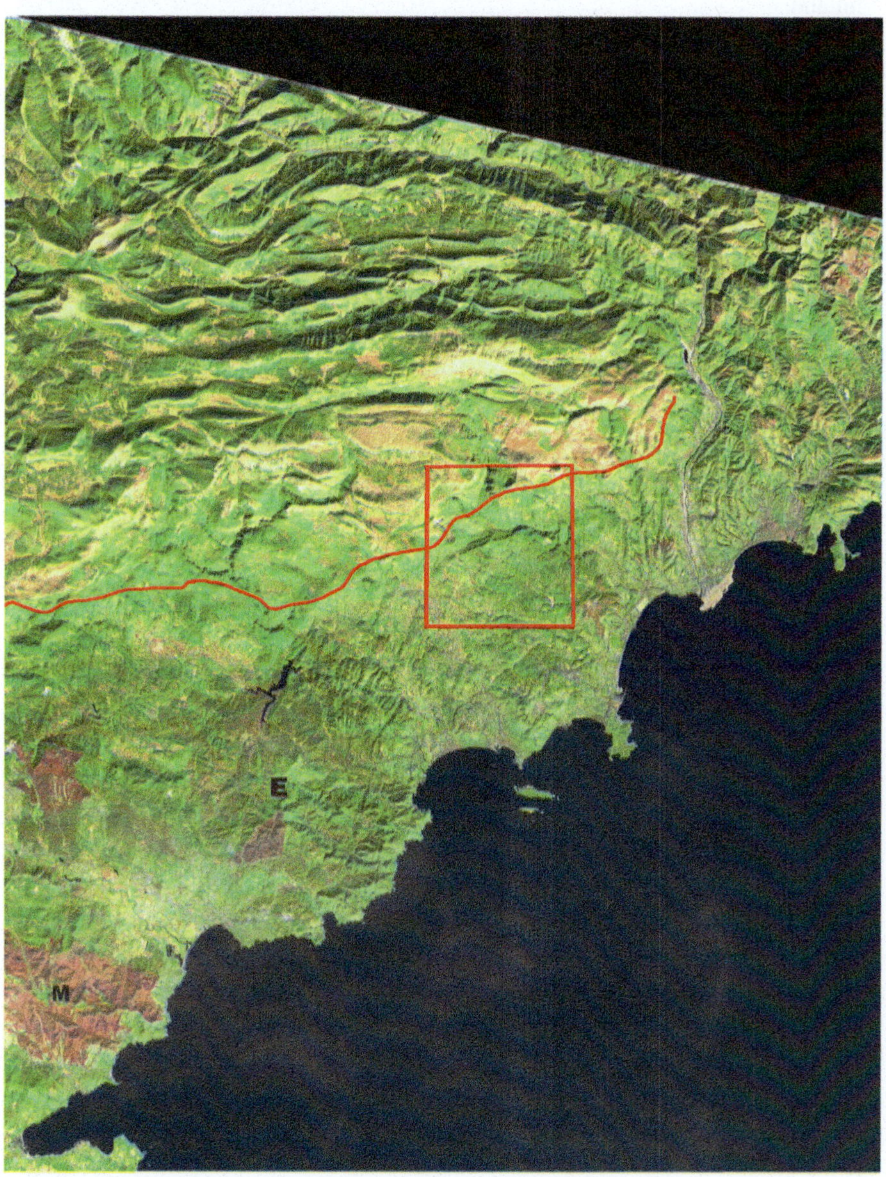

Landsat Image TM 4/5, 16 October 2003

Regional Environment

The subscene covers 2,775 km^2 of Provence in the extreme southeast of France. It is limited by 75 km of the Mediterranean coast from Menton to Fréjus, and 50 km inland.

Structure

The region is a basinal subalpine complex of polygenetic sedimentary rocks in the southern alpine structures of the Pyrenée–Provencal Tertiary Orogen. The inset frame of the site area comprises a west to east sequence of Triassic, Jurassic and Miocene foreland sedimentary units that were uplifted and selectively eroded in Mid and Upper Tertiary. The sequence is at the southern margin of a series of prominent parallel Jurassic limestone fold ridges and plateaux, the Castellane Arc. The letters **E** and **M** indicate the Esterel and Maures Hercynian massifs.

Physiography

The Mesozoic and Tertiary rocks occur as variable resistance units of carbonate, marl, detrital and shale sediments ranging from 480 to 200 m in elevation.

Apart from the city of Grasse's own suburb at the foot of the limestone plateau, the region is densely populated as residential exurbs of Cannes and Nice coastal resort cities 10 km south and eastward. Wooded areas in the tectonic area on the east are evergreen oak forests.

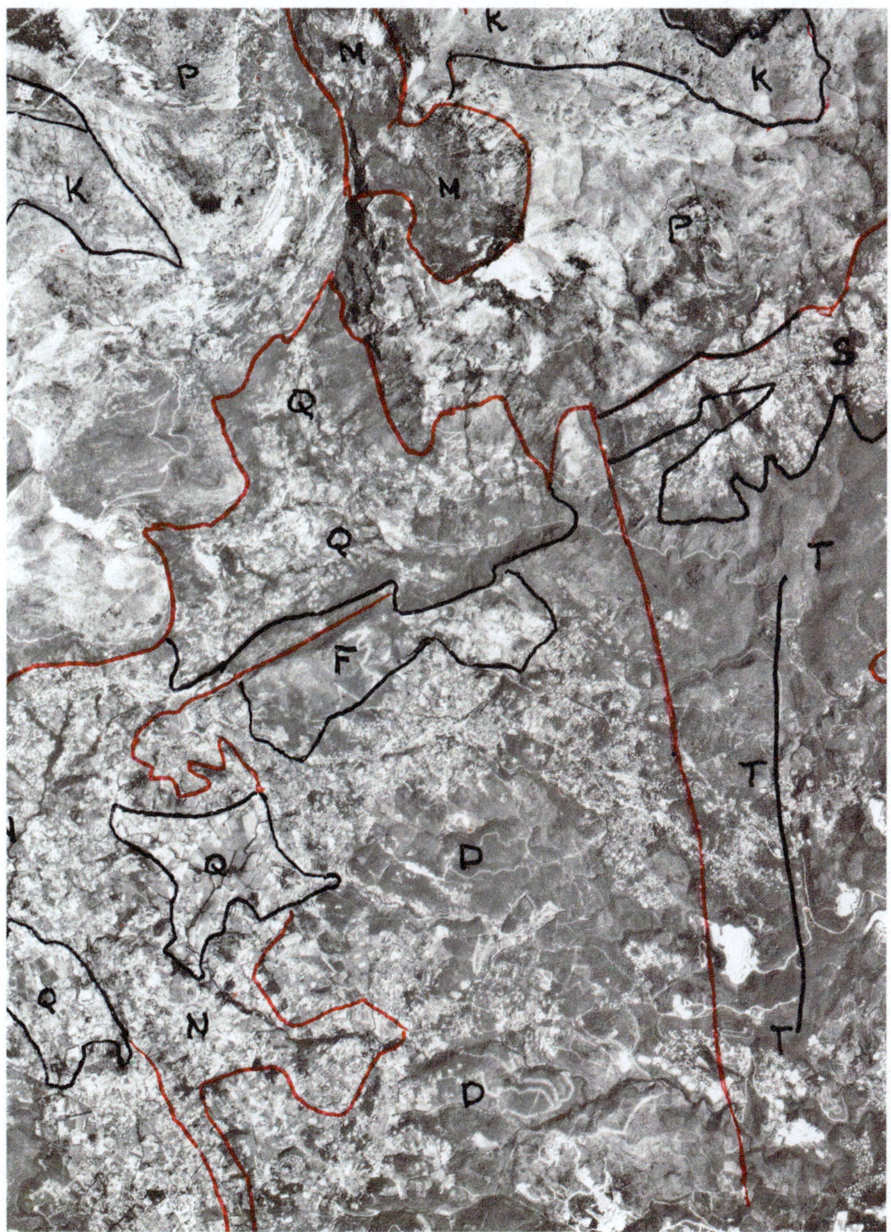

B4 Airphoto

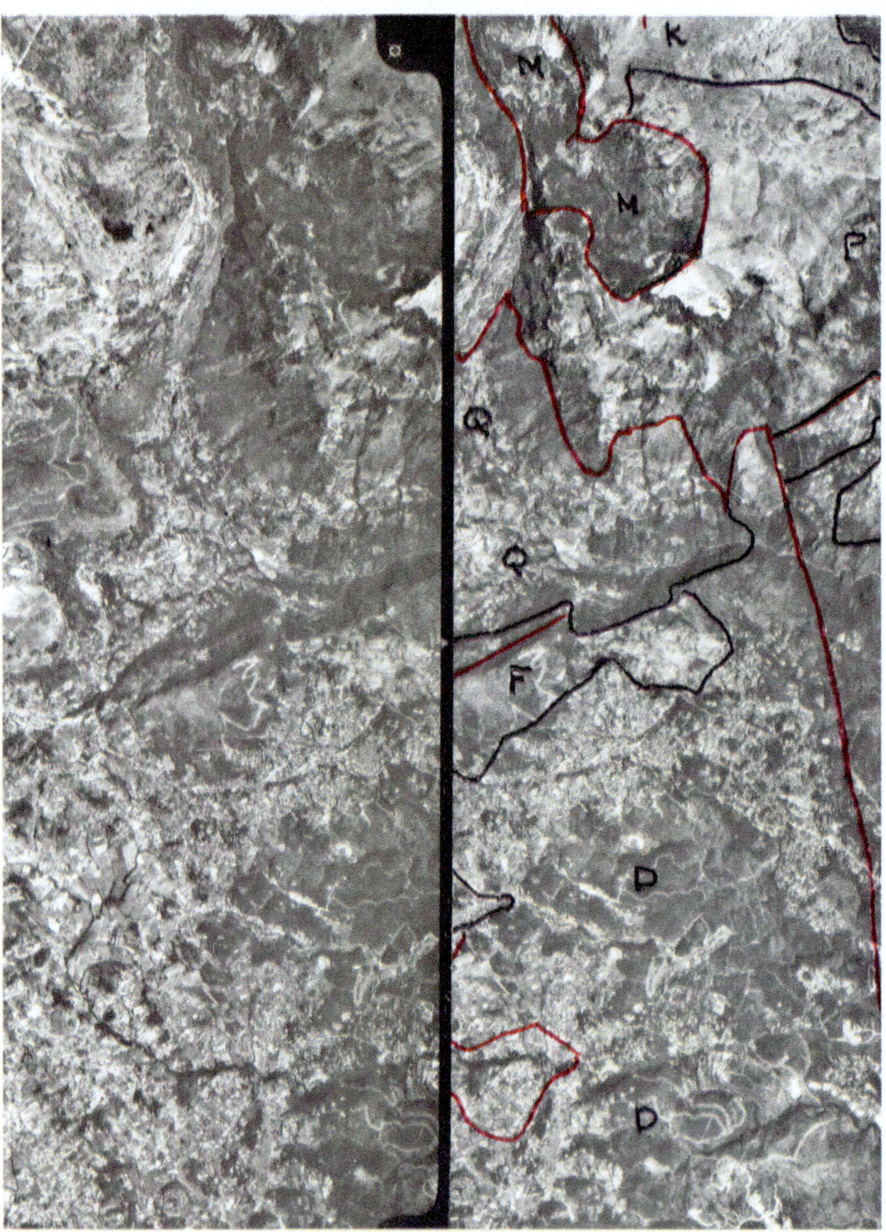

B4 Stereopair

Photo Interpretation

Nine photogeomorphic landscape units are delineated in the 1:65,000 scale, 180 km^2 framed area of the air photo.

P—The high limestone plateau. This landscape of mainly bare subtabular Jurassic surfaces at 1,300 m elevation dominates the northern third of the photo area. It occurs in two sectors divided by a landslide debris-filled canyon(M).

F—The most prominent topographic feature in the photo area, other than the plateaux, is a 1 km broad, 4 km long 480–345 m high northeast striking scarp and cuesta in the exact center of the scene. A block of Mid Jurassic limestone is raised 260 m above the large Quaternary depression on the north and 170 m above lower Jurassic dolomites on the south. The scarp is not mapped as a fault but its elevation and location suggest such a possibility. It is in a parallel sub-alignment with a mapped fault 4 km northeastward. Additionally it is part of a Landsat-resolved 20 km long arcuate feature reaching from the St-Cassien Reservoir to the southwest.

D—This polygenetic landscape is of Triassic and Jurassic strata fluvially dissected following the Upper Tertiary uplift. It occupies the central south half of the area at 200–300 m elevation.

T—The tectonic landscape is markedly faulted and occupies the east side of the area at 300 m elevation.

N—The Triassic hill landscape of interbedded dolomites and marls, is the densely populated zone of moderately sloping low hills at varying levels between the plateau and the dissected **D** landscape. The interbedding is not evident at this photo scale in the terrain.

Q—The slopes landscape is a depressed zone of steep slopes ranging from 300 to 175 m below the plateau scarps and 260 m below the topographic scarp.

S—The molasse landscape is another distinctly populated landscape at 320–300 m elevation between the foot of the plateau and the forested Tectonic landscape.

K—are two typical karst landscape zones located on the surfaces of the limestone plateaux.

M—is a debris covered wooded landslide on the scarp of the eastern plateau, descending from 1,200 to 800 m in the canyon between the plateaux.

Photo Source

Copyright IGN, 1979 7022, 17–18

References

Carte géologique de la France à 1:50,000, 1970, feuillet Grasse-Cannes.

Glintzboeckel Ch., & Horon, O. (1973). *Paysages Géologiques de Marseille à Menton*, BRGM.

Nicod, J. (1966). La Région de Cannes – Grasses, Étude géomorphologique, Recherches Régionales No 19 1966, Conseil Général des Alpes Maritimes.

Figure B5: Hérault

Location 43°32′N, 03°29′E

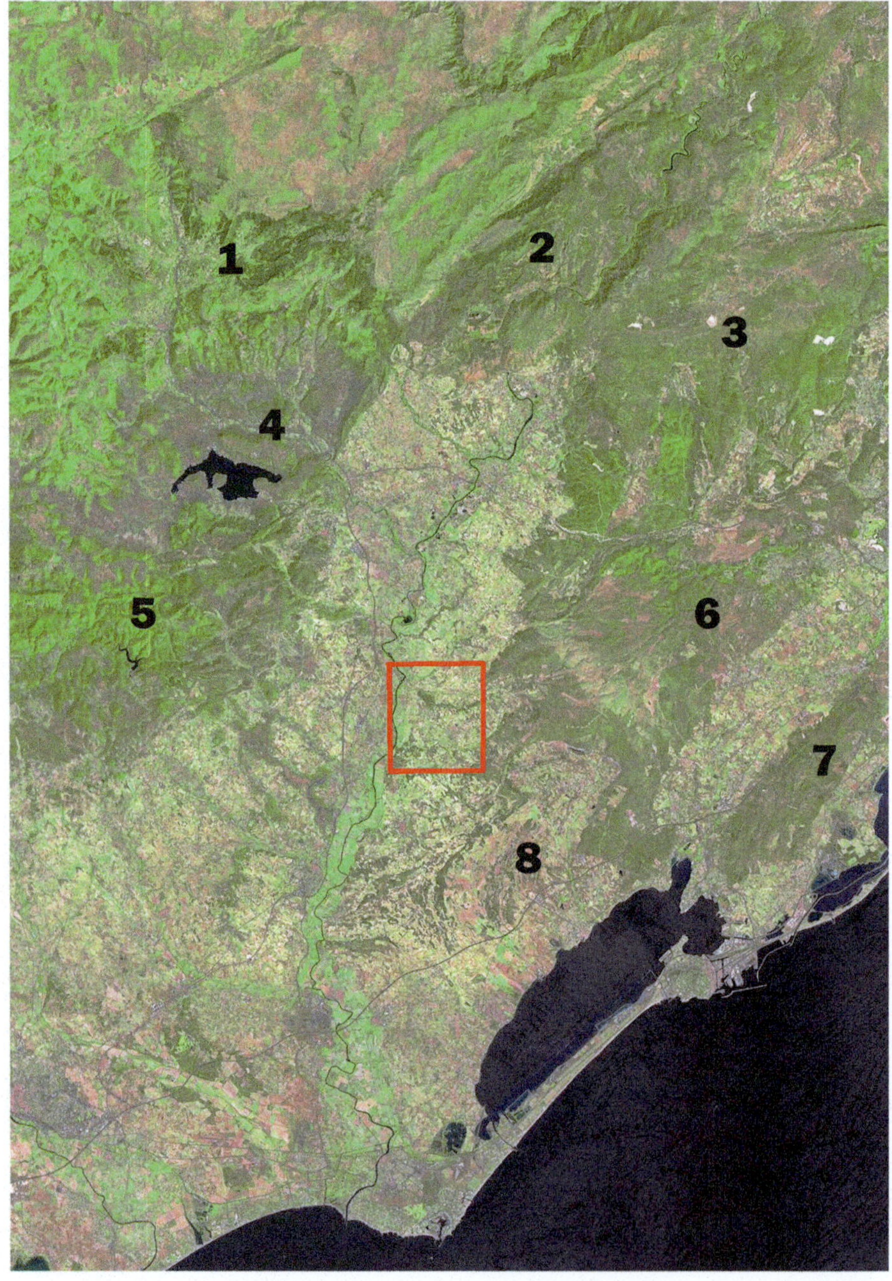

Landsat Image 4, 01 August 2011

Regional Environment

The image which is 95 km south of Figure B7 shows the site to be in the center of the Hérault structural basin that merges southward with the entirely agricultural Languedoc coastal plain of southern France.

A number of well-defined older and generally wooded regional geological units bordering the basin are identified.

1—is an extension of the *Causse* Jurassic limestone plateaus to the northwest.

2—This is a northeast striking fault zone of the limestone plateau rocks. It marks the southern end of the 150 km long macro scale fault that bounds the Cévennes zone of the Central Massif and the Rhone valley.

3—This is the western portion of a 25 km wide belt of terraced levels of Cretaceous and Eocene limestone sequences that descend to the coast.

4—The mauve area is the *Lodève* Permian basin of interbedded argillite and sandstone sediments with the *Salagou* reservoir in the center.

5—is the eastern end of the Central Massif outlier of the Precambrian *Montagne Noire*.

6—This is a unit of folded Jurassic limestones surrounded by younger sediments. The area is covered by the *garrigue* scrub of Unit K in Figure B7.

7—is the 18 km long 225 m high *Gardiole* horst of Jurassic limestone. The body of water southwest of Gardiole is the 20 km long saline *Thau* lagoon, the second largest lake in France.

8—This smallest expressed unit is an isolate 9 km long arcuate sequence of gently folded interbedded Cretaceous rocks 40–65 m high. The dark wooded crests of resistant beds define the structure.

Structure

The site is on the east bank of the narrow Hérault River in the center of a 10 km wide structural basin, an asymmetric half-graben whose downthrow on the west is the regional Unit **2** Cévennes Fault zone bordering the Central Massif. The basin is part of the upper margin of the large Golfe du Lion Basin, one of the plate tectonic Tertiary extensional basins of the western Mediterranean. The margin results from the Oligo-Miocene rifting between continental Europe and Corsica-Sardinia.

Physiography

Physiographically the Hérault Basin is the north segment of the Tertiary and Quaternary Languedoc coastal plains. It is filled with detrital and bioclastic calcareous syn-rift sediments. Except for the wooded crests of inversion hills the entire area is a monoculture of vineyards.

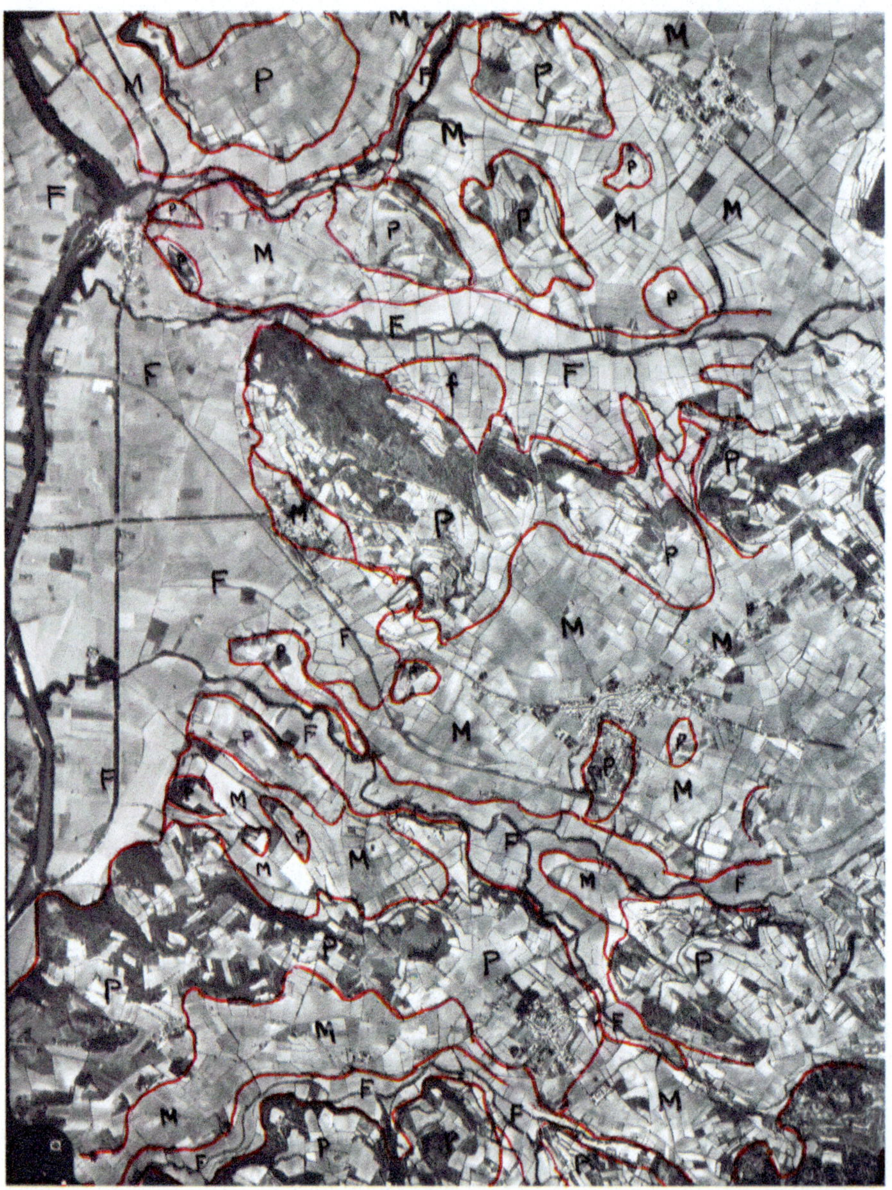

B5 Airphoto

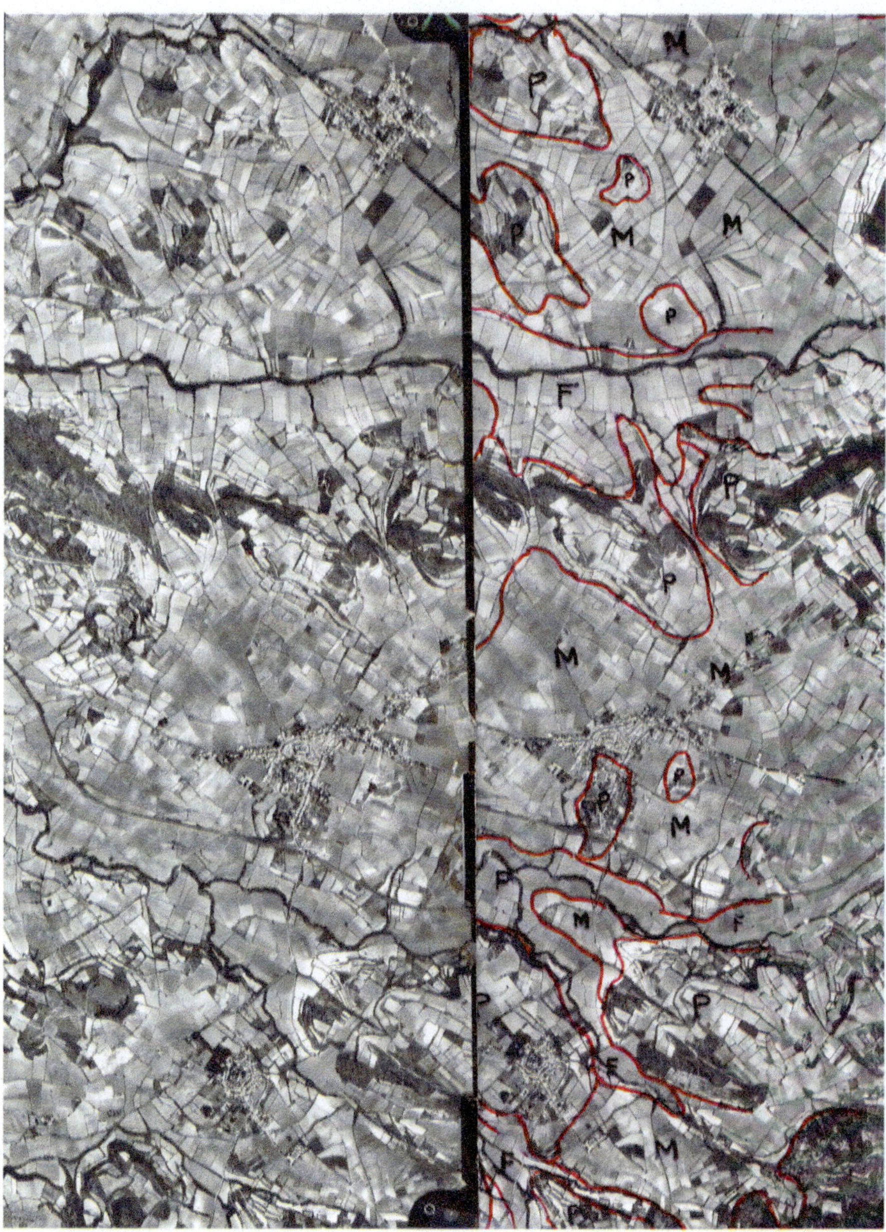

B5 Stereopair

Photo Interpretation

The portion of the 1:30,000 scale air photo covers an area of 40 km^2. Three lithostratigraphic units are distributed in 31 occurrences.

M—This is the most extensive unit. It consists of Lower Miocene marls and limestones lying as level to moderately sloping terrain from 50 to 70 m a.s.l. Three of the five local villages are located on the unit.

F—This unit comprises the 1.5 km wide, 25 m elevation low Quaternary fluvial terrace of the Hérault River and those of the three tributaries in the area.

P—This unit occurs as two prominent hills with a wooded summit rising from 75 to 115 m in elevation. They are composed of resistant stream channel fill of Pliocene conglomerates now in inverted relief.

Photo Source

Copyright IGN, 1974 2544, 4–5

References

Le bassin Oligo-Miocène de l'Hérault: un example de rétrocharriage des structures pyrénéennes Implications hydrogéologiques. Rapport final, BRGM/RP–53733–FR, janvier 2005.

Séranne, M., Benedicto, A., Labaum, P., Truffert, C., & Pascal, G. (1995). Structural style and evolution of the Gulf of Lion Oligo-Miocene rifting: Role of the Pyrenean Orogeny. *Marine and Petroleum Geology, 12*(8), 809–820.

Figure B6: Toulon

Location 43°16′N, 05°55′E

Landsat Image T/M 4-5, 16 October 2003

Regional Environment

The area is in western Provence, 45 km east of Marseille, 10 km north of Toulon, and bounded eastward by the coastal Hercynian Maures Massif.

Structure

The site is in the eastern part of the structural unit *Beausset*, a synclinal basin, in a region of east-west striking Mesozoic limestone folds the Pyrenée-Provencal Tertiary Orogen.

Physiography

The country of western Provence is one of considerable complexity. In the Beausset area differential erosion resulting from the Tertiary Orogen has etched out hard rocks into rolling hilly relief, trenched by superimposed gorges and linked by broad shallow valleys in softer rocks.

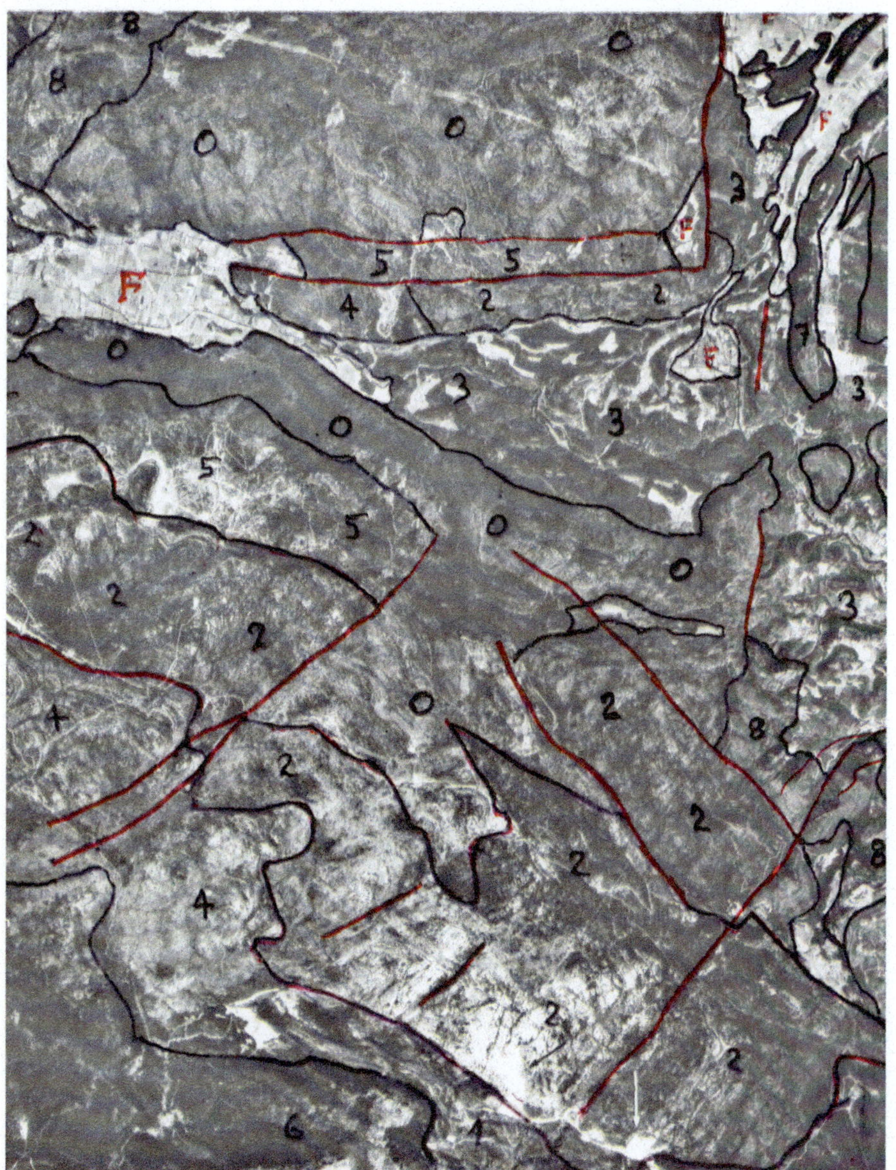

B6 Airphoto

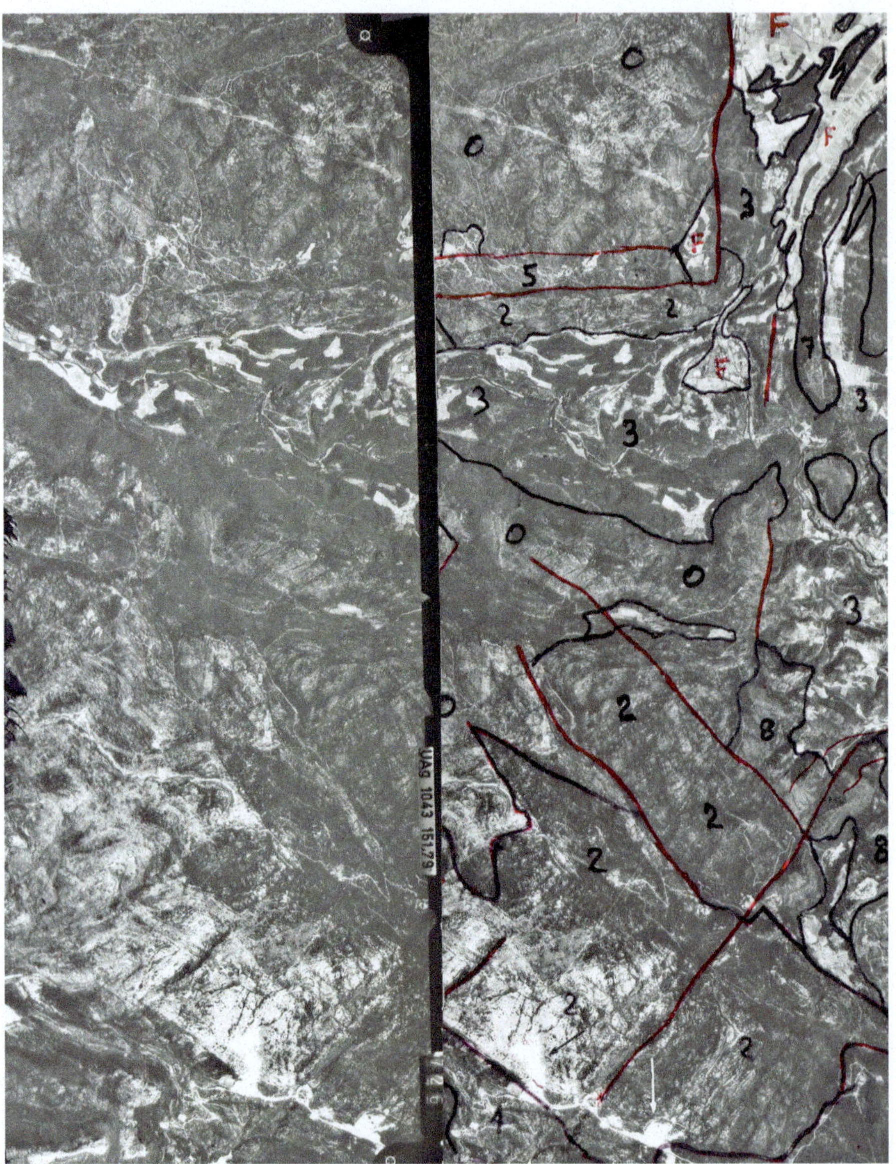

B6 Stereopair

Photo Interpretation

The 180 km^2 1:65,000 scale photo area consists of a complex of nine units of individually interbedded Mesozoic sedimentary rocks ranging in elevation from 200 to 850 m. Apart from two limited agricultural enclaves the entire area is covered in a low woodland and scrub vegetation (garrigue) of evergreen oak.

0—locates the most extensive and, at 800–850 m elevation, the highest unit in the area. It is composed of relatively little disturbed Upper Jurassic dolomite with some visible interbedding of limestone breccia. Occurrences in the north and center are massive-appearing but well dissected.

2—This unit at 700 m elevation delineates a 3 km broad and 10 km long segment of a 55 km long Lower Cretaceous arcuate reef and bioclastic limestone that extends from here to Marseille. This unit and unit 4 constitute a sequence of reefal sediments that form the southwest dipping margin of the regional *Beausset* synclinal basin visible on the Landsat image. Bauxite quarries are visible along the south side of the sequence.

The unit is crossed by northeast and northwest striking faults, the westerly one being a strike-slip fault.

3—These occurrences at 250–350 m elevation are downfaulted and dissected interbedded weak Triassic argillaceous limestone and marl.

4—At an average 450 m elevation this unit consists of well interbedded Upper Cretaceous reef limestones, argillaceous limestone and marl.

5—This 700 m elevation 4 km long faulted ridge of Lower Cretaceous interbeds of marly argillaceous and thick limestones is a reef-flank unit of Unit **2** lying above the Jurassic and Triassic lowlands.

6—This unit of limited occurrence on the air photo is at the eastern limit of the 300 km^2 *Beausset* basin that is bounded on the north by the reef Units **2** and **4**. The basin is the described major structural unit of western Provence. At elevations ranging from 200 to 450 m it consists of interbeds of marl, marly sandstone and reefal limestone.

7—These are 200–220 m elevation low dipping clearly interbedded Mid-Jurassic marls and argillaceous limestone.

8—This faulted minor unit at elevation 400–450 m consists of well interbedded Upper Cretaceous limestone and marl.

F—These are agriculturally occupied alluvium filled poljes associated with the downfaulted Unit **3**.

Photo Source

Copyright IGN, 1979, 7023, 26-27

Reference

Rouire, J. (1979). Notice explicative de la feuille Marseille. *Carte Géologique de la France à* 1; 250,000, BRGM

Figure B7: Vivarais

Location 44°24′N, 04°09′E

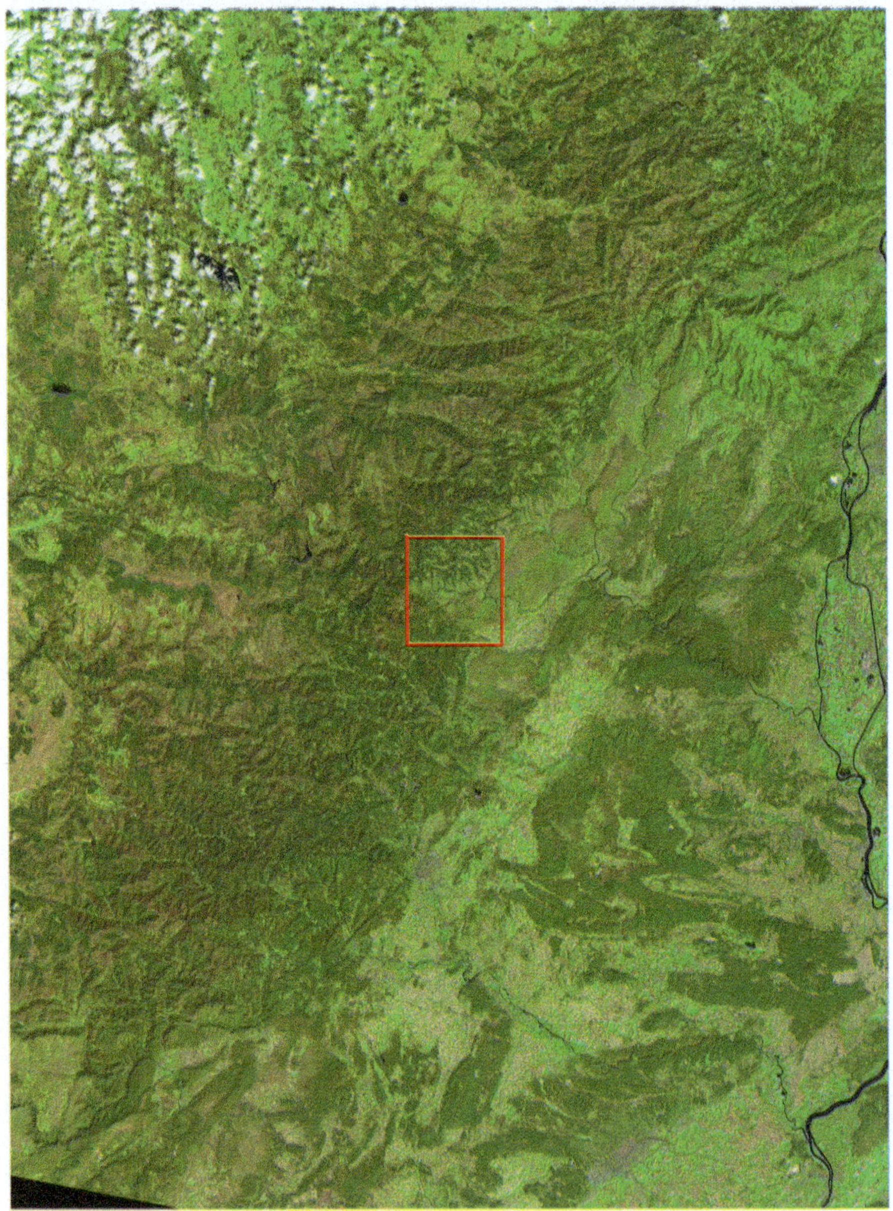

Landsat Image T/M 4-5, 11 April 2011

Regional Environmemt

The scene is 95 km north of Figure B5. The western half is in the south-eastern edge of the Massif Central known as Cévennes. The eastern half is a 35 km wide belt of folded and faulted Mesozoic sedimentary rocks that lie between the Cévennes and the lower Rhone Valley. The Example site is in the inner margin of the sedimentary belt.

Structure

The area is at the southern extremity of a northeast striking 150 km long belt of Jurassic and Triassic sediments that are in faulted contact with the southeastern margin of the French Central Massif. A northwest striking fault that offsets the sediments in the photo area marks the limit of the sedimentary belt.

Physiography

The Jurassic sediments are plateaux of karstic thick limestones at average 260 m elevation; the Triassic rocks are dissected interbedded sandstones, shaly limestones and marls ranging from 200 to 350 m elevation.

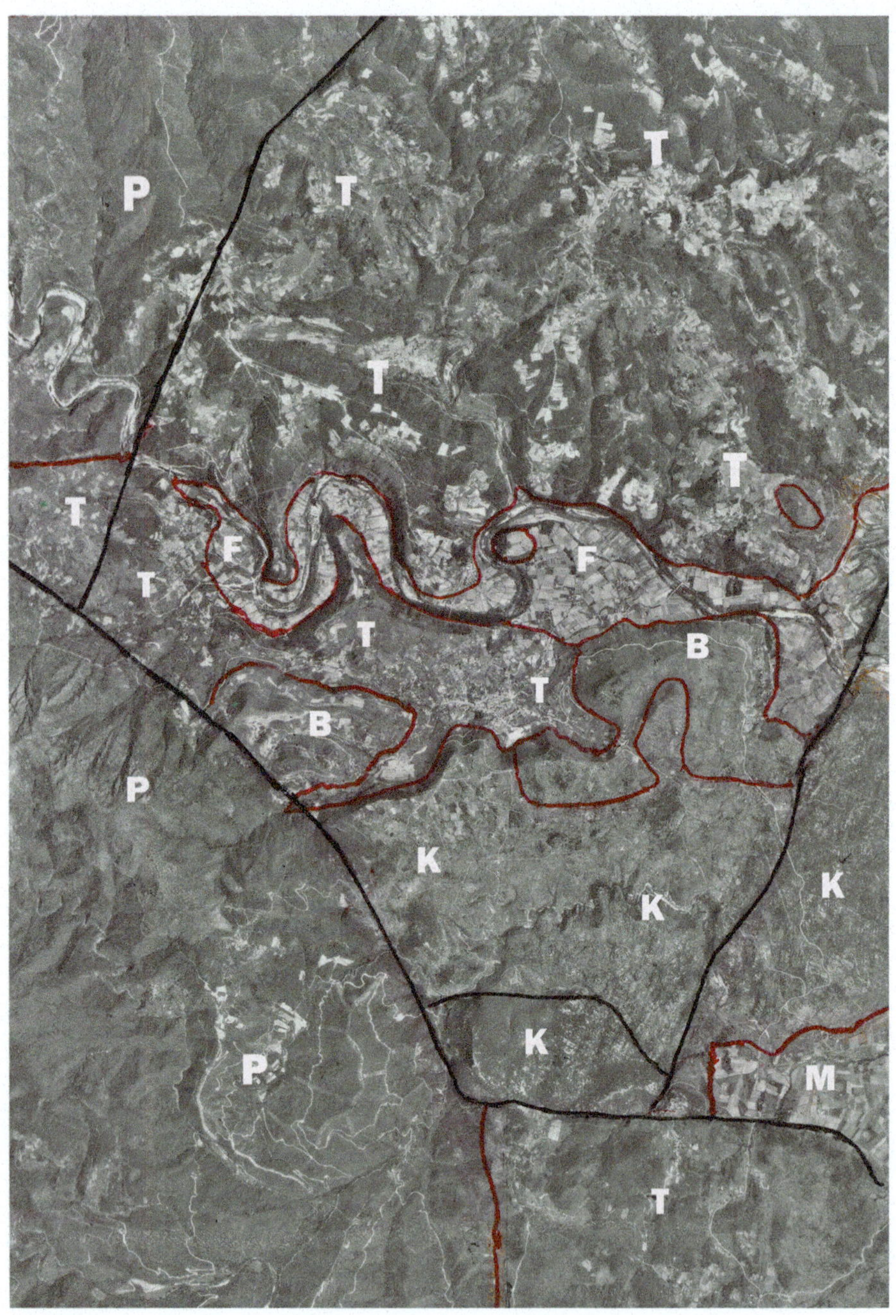

B7 Airphoto

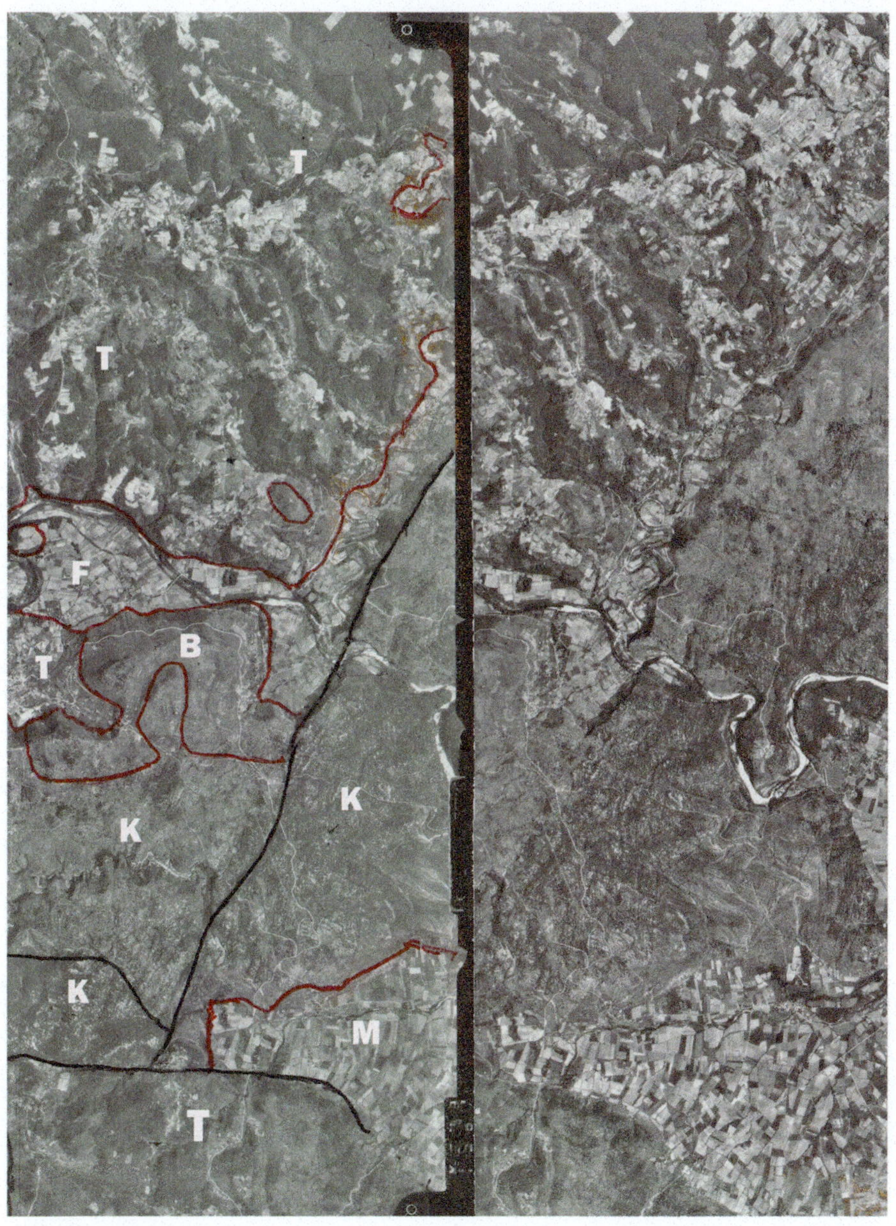

B7 Stereopair

Photo Interpretation

The 1:60,000 scale air photo area is 125 km^2 with 12 occurrences of six lithologic units.

F—These are the cultivated terraces of the Chassezac River 175 m elevation upstream to 145 m downstream.

M—at 145 m elevation this is the cultivated local lowland unit of Lower Cretaceous marl and marly limestone.

K—This is the southern unit of the Upper Jurassic limestone belt. It occurs in three sectors in the photo area. The central sector is at 230–280 m elevation and is incised by a stream leading to the agricultural **M** unit lowland. The darker sector at the same elevation east of the fault that divides the Jurassic and Triassic rocks northeastward off the photo has a denser cover of *garrigue* scrub. The small sector is a 1,200 by 2,400 m block-faulted cuesta raised 170 m above the other sectors.

B—This unit of two occurrences at 275–320 m elevation of Lower Jurassic interbedded marly limestone, sandstone and dolomite lies above and between the limestone plateau and the Triassic lower ground.

T—This largest unit in the photo at 200–350 m elevation area regionally parallels the limestone plateau. It consists of Triassic interbedded shaly limestone, sandstone and marl. These relatively weak sediments are deeply fluvially dissected with cultivable land restricted to the narrow interfluves.

P—These rugged hills of gneiss and schist of the Paleozoic Hercynian Central Massif are in faulted contact with the Mesozoic sedimentary units of the photo area and the region extended northeastward. The hills are fairly densely forested with oak, beech and chestnut at lower elevations.

Photo Source

Copyright IGN, 1978, 7022, 55-56

References

Clerc, C. (2009). Structure et fontionnement du système karstique de Saint-André-de-Cruzières, Université Montpellier2 et SupAgro Montpellier.

Laruelle, S. (1995). La vallée du Chassezac, étude géomorphologique, Mémoire de Géographie, Paris.

Figure B8: Port de Paix

Location 19°50′N, 72°50′W

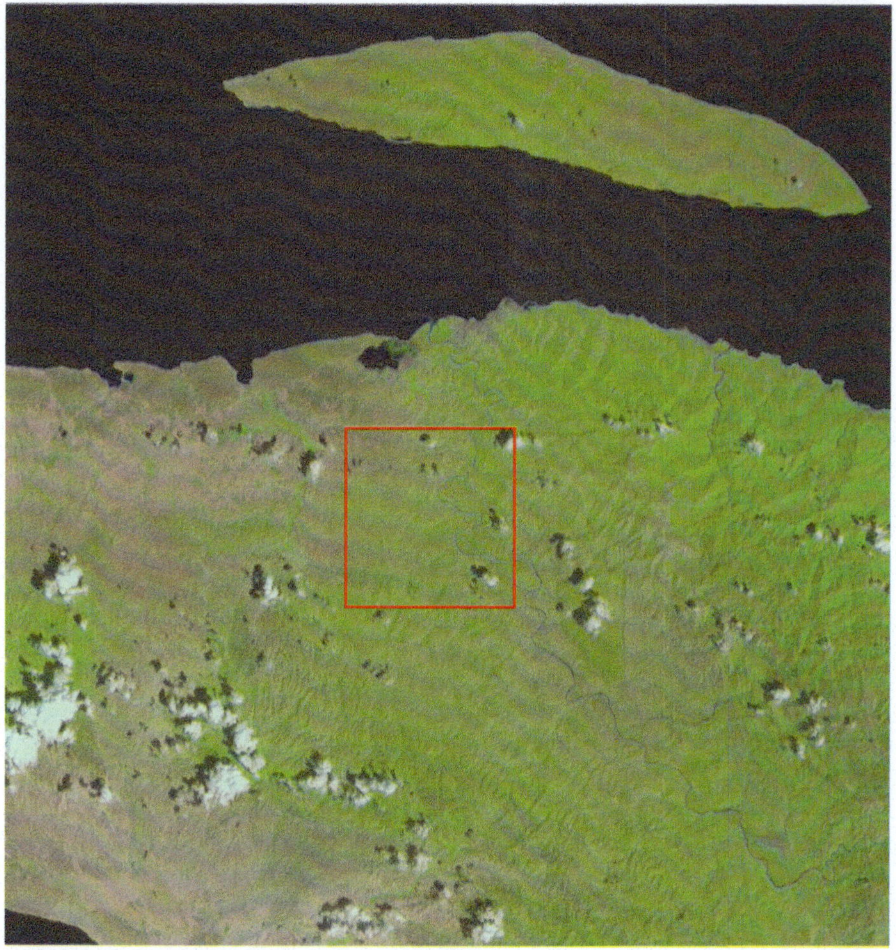

Landsat Image T/M 4-5, 18 September 2007

Regional Environment

The Example is located in the Northwest Peninsula of Haiti.

Structure

Haiti Island is part of the Greater Antilles Deformed Belt, of folded Mesozoic and Cenozoic sedimentary rocks. The belt is along the strike-slip boundary of the North American and Caribbean tectonic plates.

Physiography

The Example area is located 10 km inland from the northwest coast in a 7 km wide fault-bounded trough, not clearly resolved in the image, that separates the northwest peninsula from the northern coast eastward.

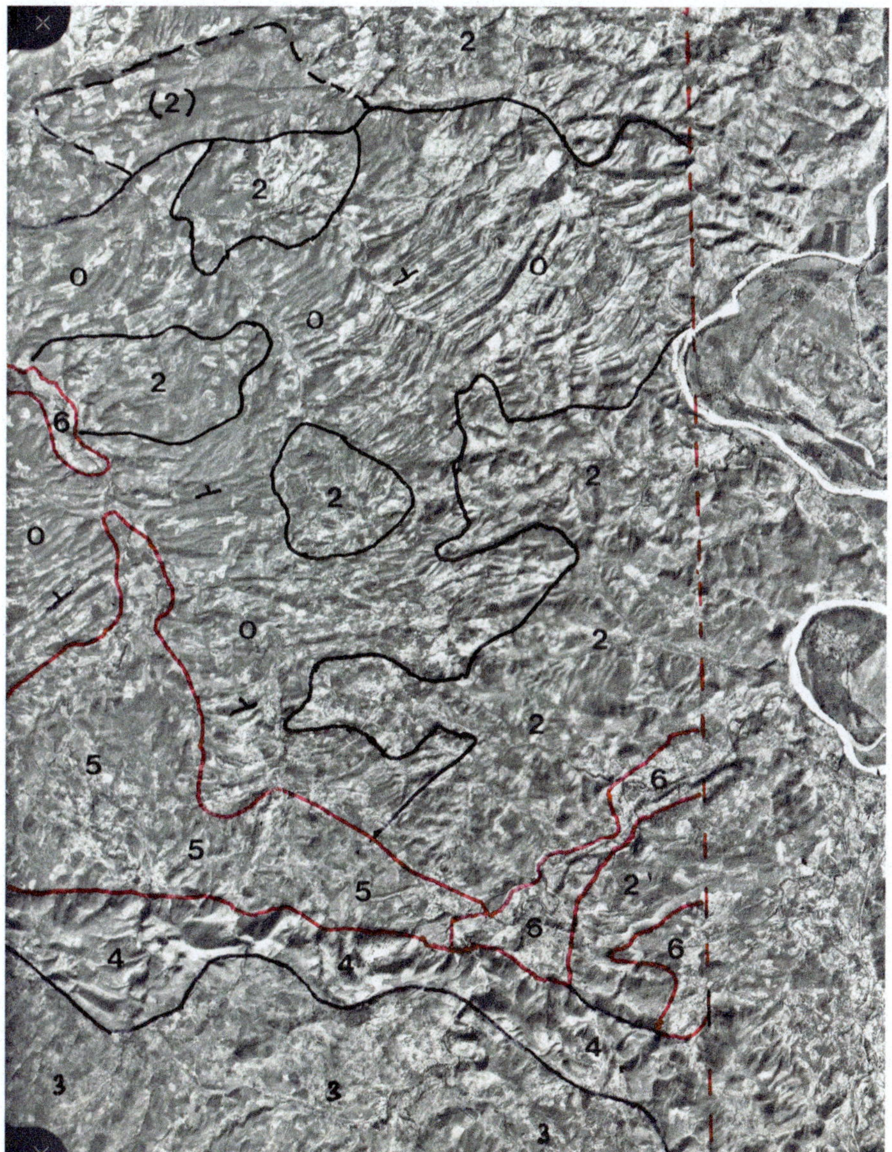

B8 Airphoto

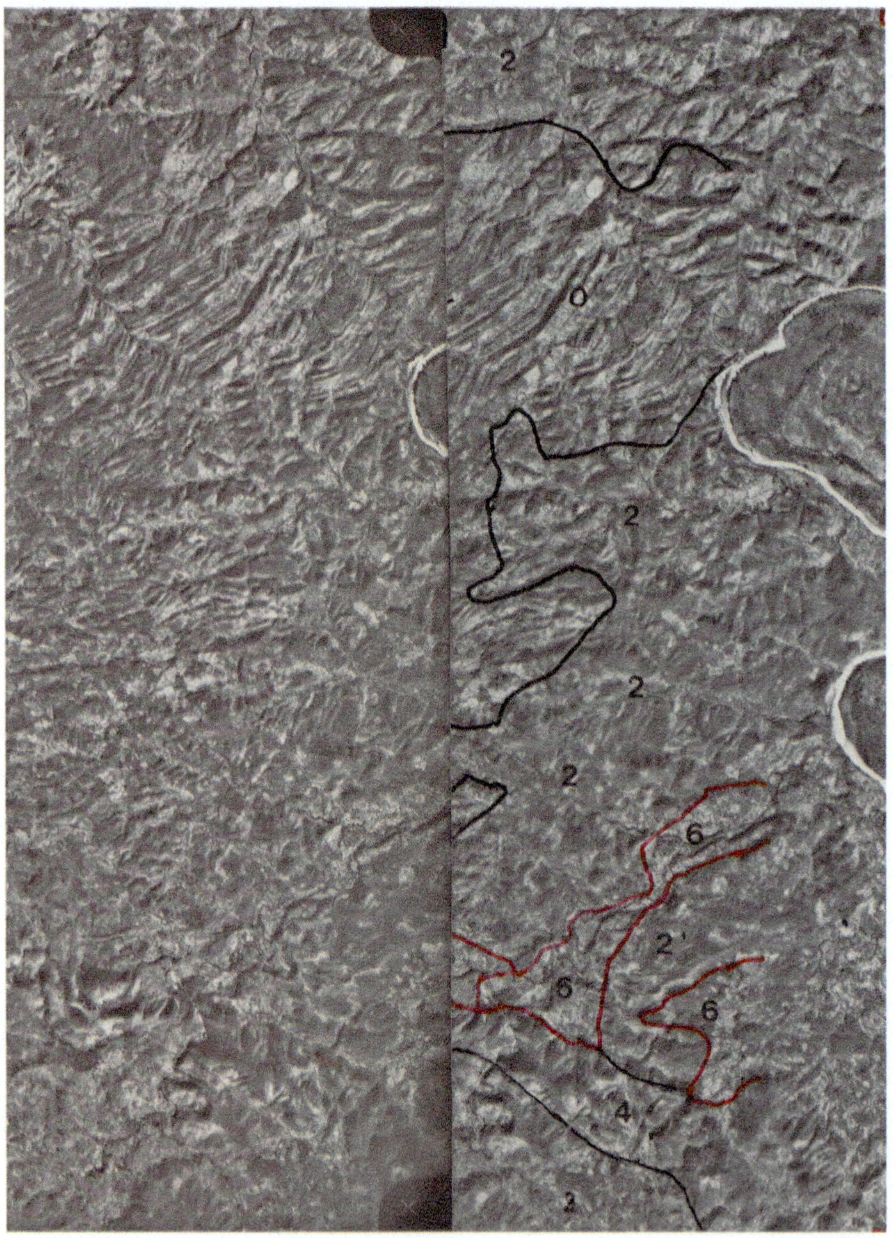

B8 Stereopair

Photo Interpretation

The 1:40,000 scale air photo area is 45 km^2. There are thirteen occurrences of six lithologic units.

O—These are thinly interbedded northwest dipping Lower Miocene sandy shales, limestones and marls at elevations ranging from 150 to 200 m.

2—These are strongly dissected, horizontally bedded Oligocene limestones, marls, sandstones and shales 100–150 m elevation.

3—This a unit of dissected Eocene limestone shale and conglomerate.

4—This unit is the north end of an arc shaped isolated 80 m high Eocene limestone ridge.

5—This unit is interpreted as a poorly consolidated intermont basin of Pliocene alluvial sediments.

6—These are alluvial terraces of a short stream valley.

Photo Source

Copyright IGN, 1978, -HAI-400, 109-110

References

Butterlin, J. (1960). *La Géologie de la République d'Haïti et ses rapports avec celle des régions voisines*. Mémoires de l'institut Français d'Haïti.

Inventaire des ressources minières de la République d'Haïti, Dossier promotionnel Fascicule 1 1990, Bureau des Mines et Énergie.

Notice explicative de la carte géologique d'Haïti au 1:250,000, 1988, BRGM/BEICIP/BME.

Woodring, W. P., Brown, J. S., & Burbank, W. S. (1924). Geology of the République of Haiti. *Geological Survey of the Republic of Haiti*.

Figure B9: Edziza

Location 57°39′N, 130°38′W

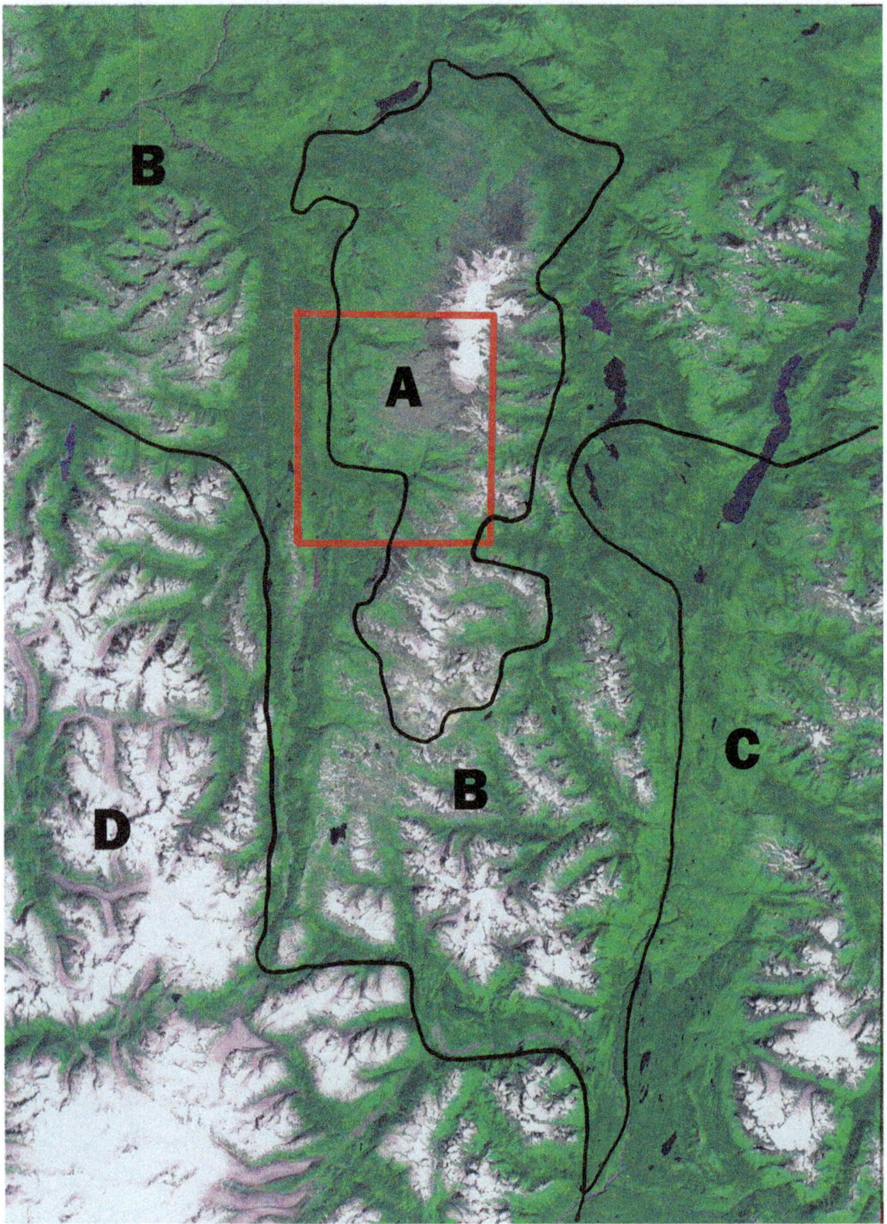

Landsat Image MSS, 09 August 1986

Regional Environment

The scene covers 10,500 km^2 of the Canadian Cordillera in north central British Columbia. The isolated ice cap of the Edziza Volcano stands above surrounding valleys.

Structure

The Mount Edziza Volcanic Complex (MEVC) is part of the Stikine Volcanic Belt, a broad belt of Miocene and younger volcanoes which runs roughly parallel to the continental margin through northwestern British Columbia and southern Yukon.

Physiography

Four major units occur within the scene area.

Unit **A** is the MEVC

Unit **B** is an intermontane belt. Upper Paleozoic sedimentary and volcanic rocks in the belt are tightly folded along north south axes. A pluton of dissected Mesozoic granite occurs below code **B** in upper left.

Unit **C** are Mesozoic sediments of the Skeena Mountains.

Unit **D** are ice fields of the Coast Mountains with elevations above 2,000 m.

Physical features within MEVC vary according to the type of material erupted and the surface on which it was deposited. Because of its low viscosity basaltic lava of Unit **L** of the air photo is spread extensively in thin flows. The flows are locally covered by tephra and glacial deposits. The north-south axis of the volcanic complex comprises four composite volcanoes in various stages of dissection. Mount Edziza lies just north of the north edge of the air photo. It rises to a circular ice-filled crater about 2 km in diameter at 2,600 m elevation.

Alpine glaciers and an ice cap (**G1** on air photo) mantle most of the upper cone down to an elevation of 2,100 m. The largest glacier is Tenchen (**gt** on air photo).

Figure E6 is located at the **B** code in the northwest.

R
R
8
7
L
P
L
gt
L
7
7
V
L
F
L
P
R
L
L
8
P
8
L
6
18
P
L
L
8
R
L
6
V
F
7
L
L
f
F
5
R
15
R
7
7
F
R
F
L
L
L
5
F
F
L
F
L

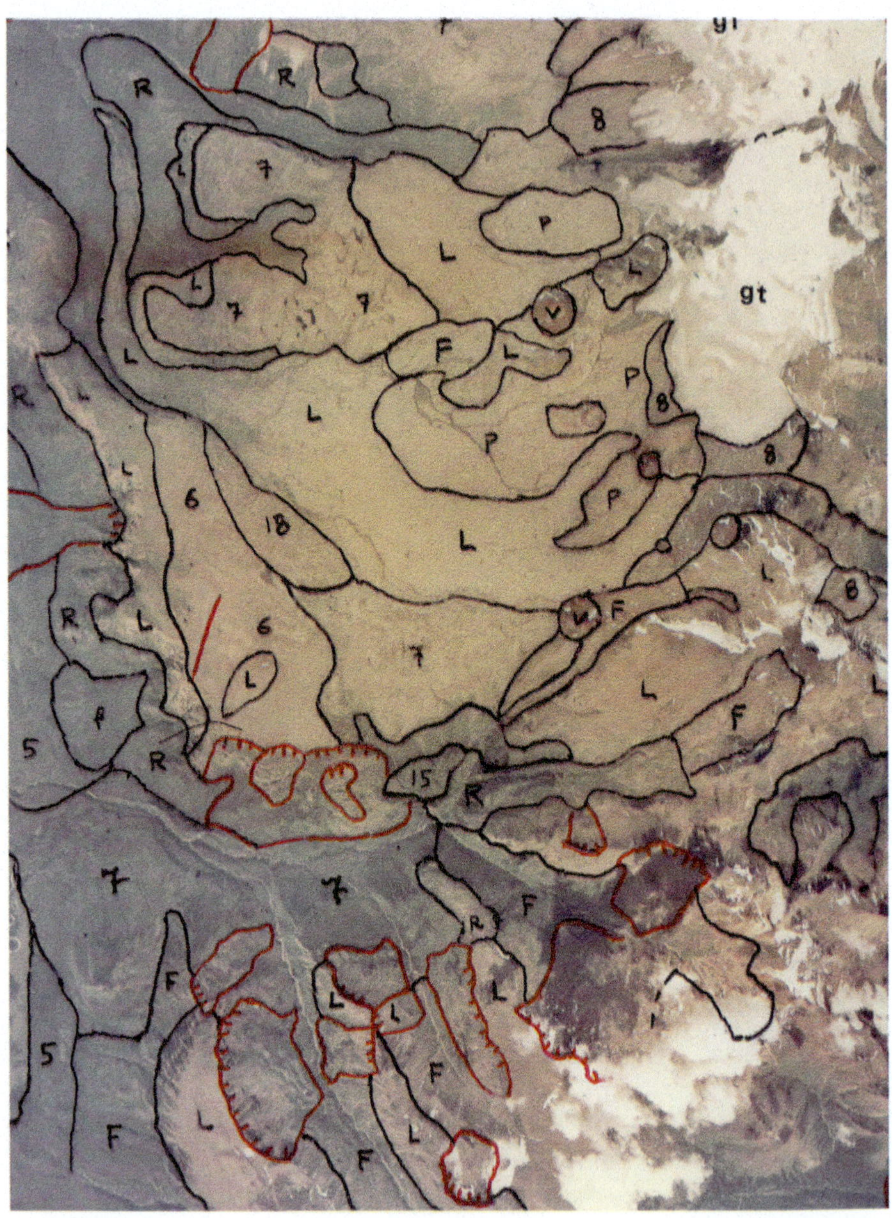

B9 Airphoto

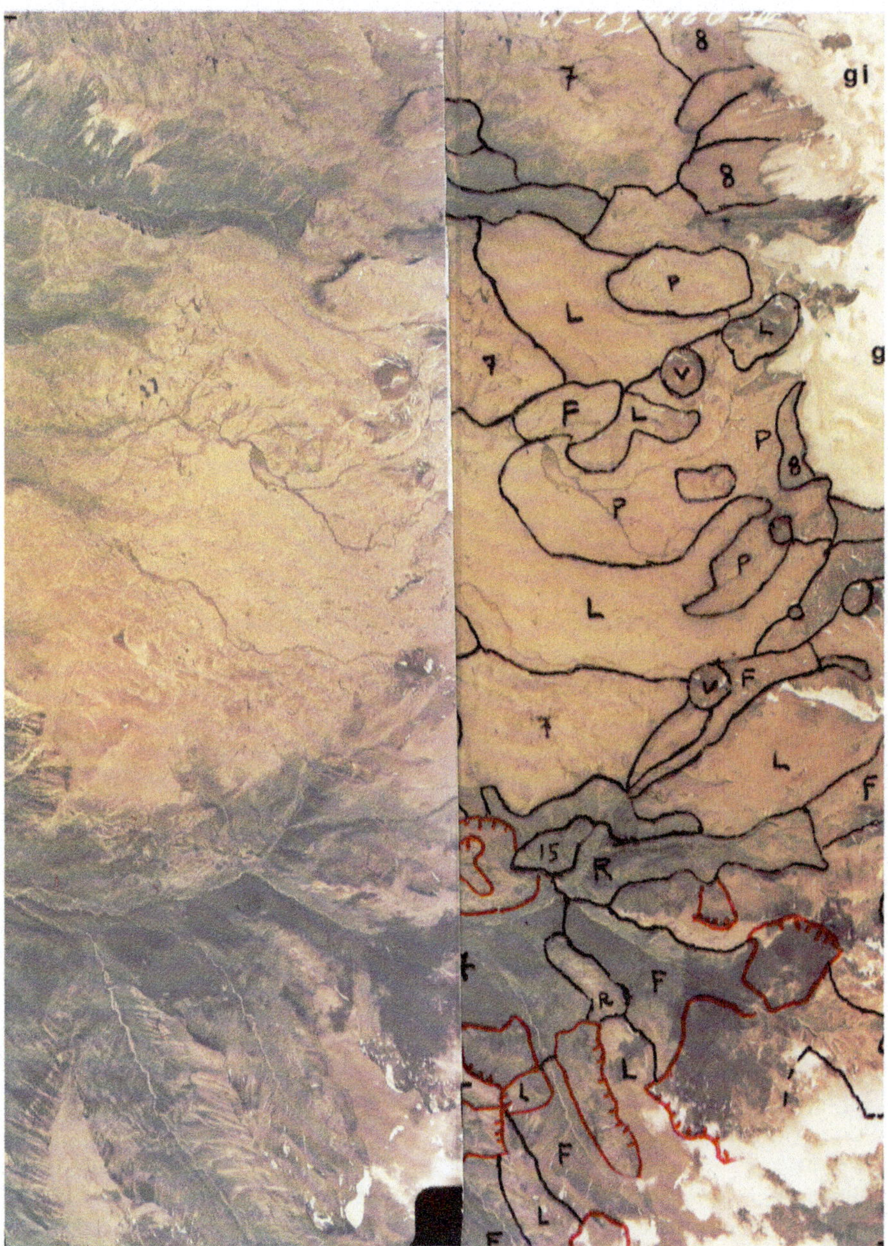

B9 Stereopair

Photo Interpretation

The 1:125,000 scale colour air photo is 17 by 22 km and 375 km^2 in area.

Three units are volcanic; they postdate the last episode of regional glaciations.

L—These are thick sheets of lava flows on the gently sloping surface of the surrounding plateau. They issued from a cluster of vents which lie near the lower edge of Glacier **Gt** (Tenchen) at 1,800 m elevation. The main body of the lava field flowed in a thick tongue a distance 15 km with a vertical drop of 840 m to 960 m. The narrow unit on the left is the steep lava escarpment that marks the western boundary of the plateau.

V—is a 3 km broad 130 m high pyroclastic cone (Cocoa).

P—is unconsolidated airfall pumice that was deposited on the lava flows.

Bedrock Unit

R—is a 1,350 m high mass of Paleozoic schist and gneiss on the west boundary of the volcanic plateau.

Glacial Deposits

5—is reworked moraine on a bench above Mess Creek.

6—is a veneer of ground moraine over the western part of the main lava field.

7—are occurrences of hummocky moraine which are marginal to and 30–40 m above lava flows.

8—These are marginal moraines within the trimlines of the **Gt** glacier.

Surficial Deposits

F—are talus on and at base of gully slopes at the south end of the photo area

f—This is a 2.5 km wide and 1.5 km long alluvial fan lying 600 m below the western escarpment of the lava plateau. The fan descends 275 m from 1,065 to 790 m.

—Red areas are landslide deposits resulting from receding ice leaving unstable oversteepened valley walls which collapsed in a series of landslides in the southern part of the air photo area.

Photo Source

Courtesy of National Air Photo Library, RSA30552, 12-13

Reference

Souther, J. G. (1992). The Late Cenozoic Mount Edziza Volcanic Complex, British Columbia, Geological Survey of Canada Memoir 420.

Figure B10: Aullène

Location 41°48′ 25″N, 09°05′16″E

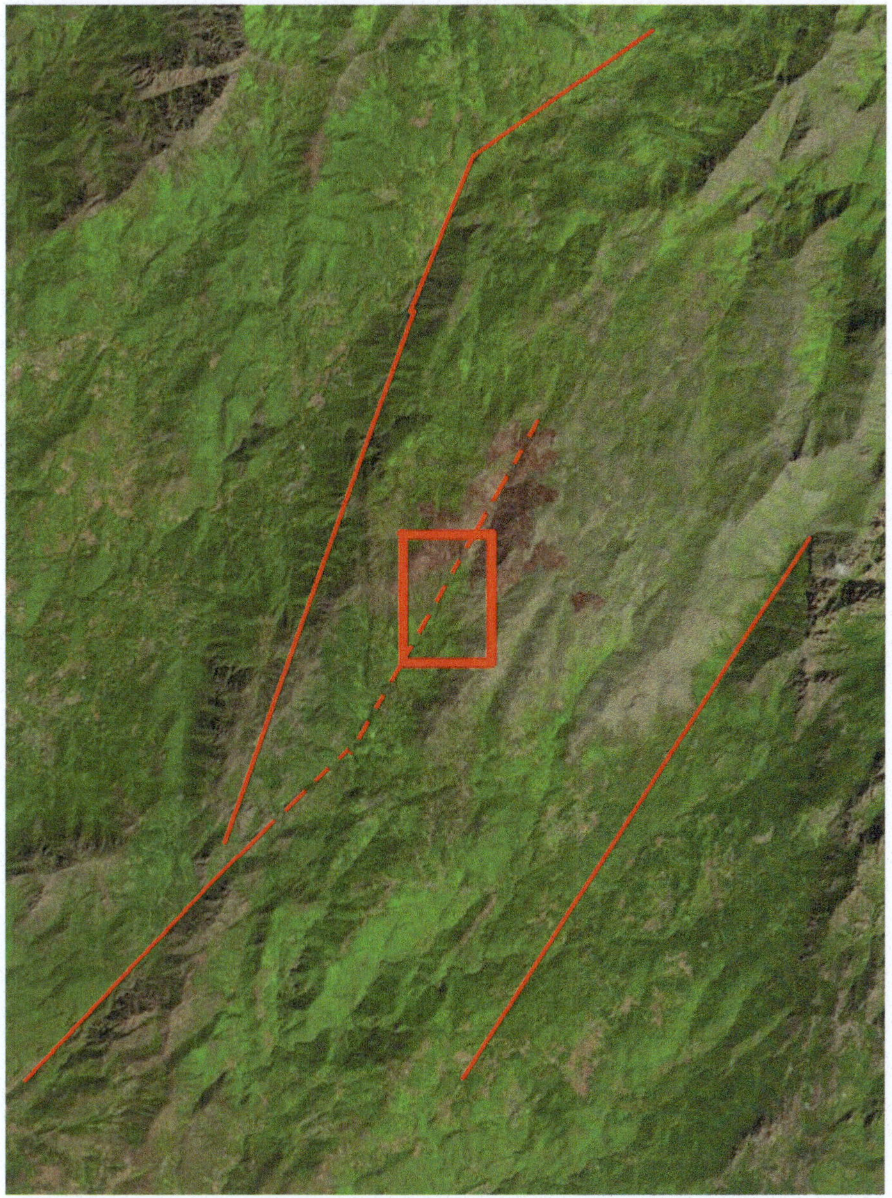

Landsat Image T/M 4-5, 16 September 2003

Regional Environment

The image covers 650 km^2 of the granitoid Hercynian terrain of south central Corsica. The imaged portions of three mapped major strike-slip faults have been drawn. The broken line traced through the Example site is extended from one of the known faults along a visible lineament.

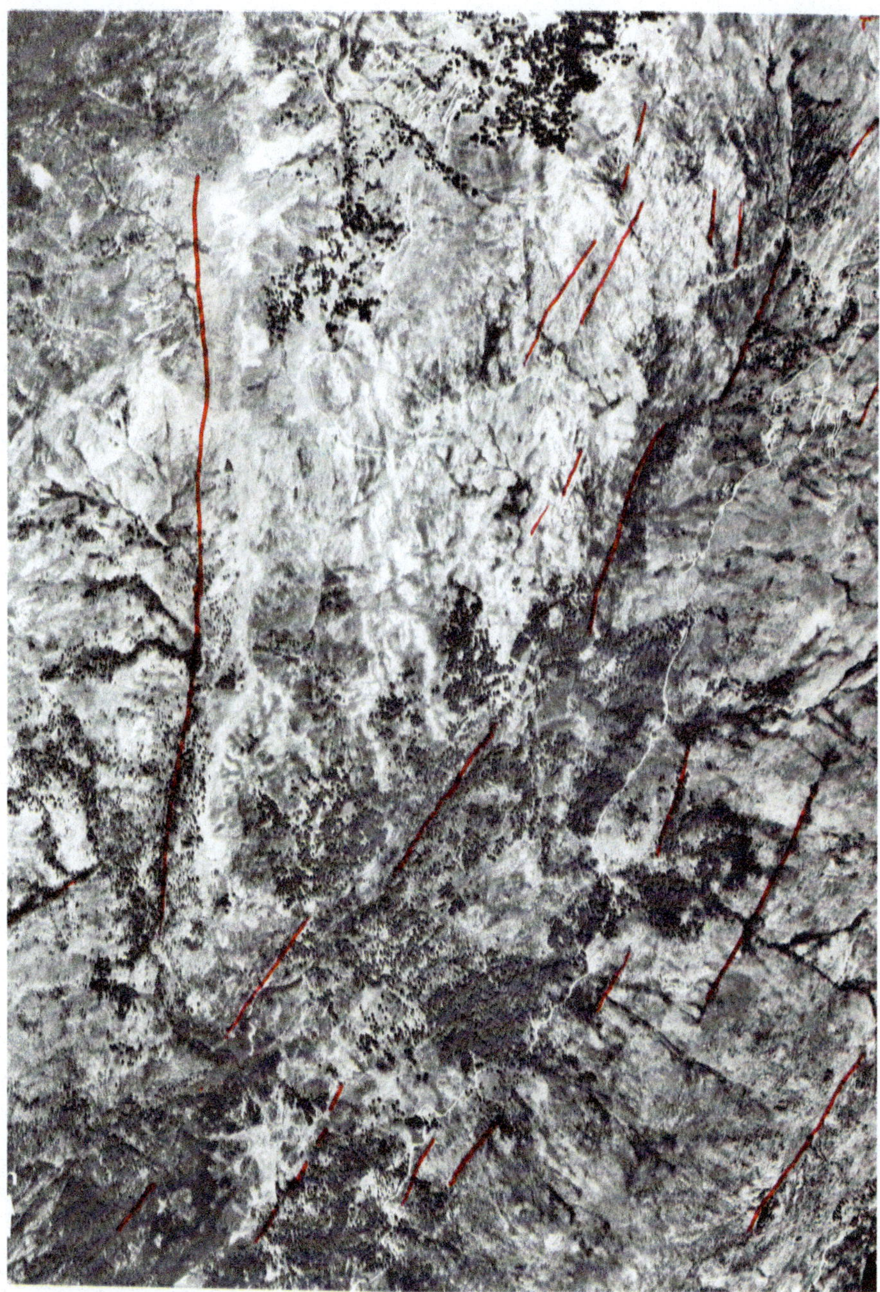

B10 Airphoto

B10 Stereopair

Photo Interpretation

Fracture traces are drawn on this 1:25,000 scale air photo in a longitudinal depression at 880 m elevation in an area of rugged ground at elevations ranging between 1,100 and 1,200 m. They are parallel to the strike of one of the mapped fault line traces and are interpreted as a possible extensional splay from the fault.

Photo Source

Copyright IGN, 1974, 4253, 101-102

Reference

Bouchra Zarki-Jakni, et al. (2004). Cenozoic denudation of Corsica in response to Ligurian and Tyrrhenian extension: results from apatite fission-track thermochronology. *Tectonics 23*.

Figure B11: Sospel

Coordinates 43°50′29″N, 07°26′17″E

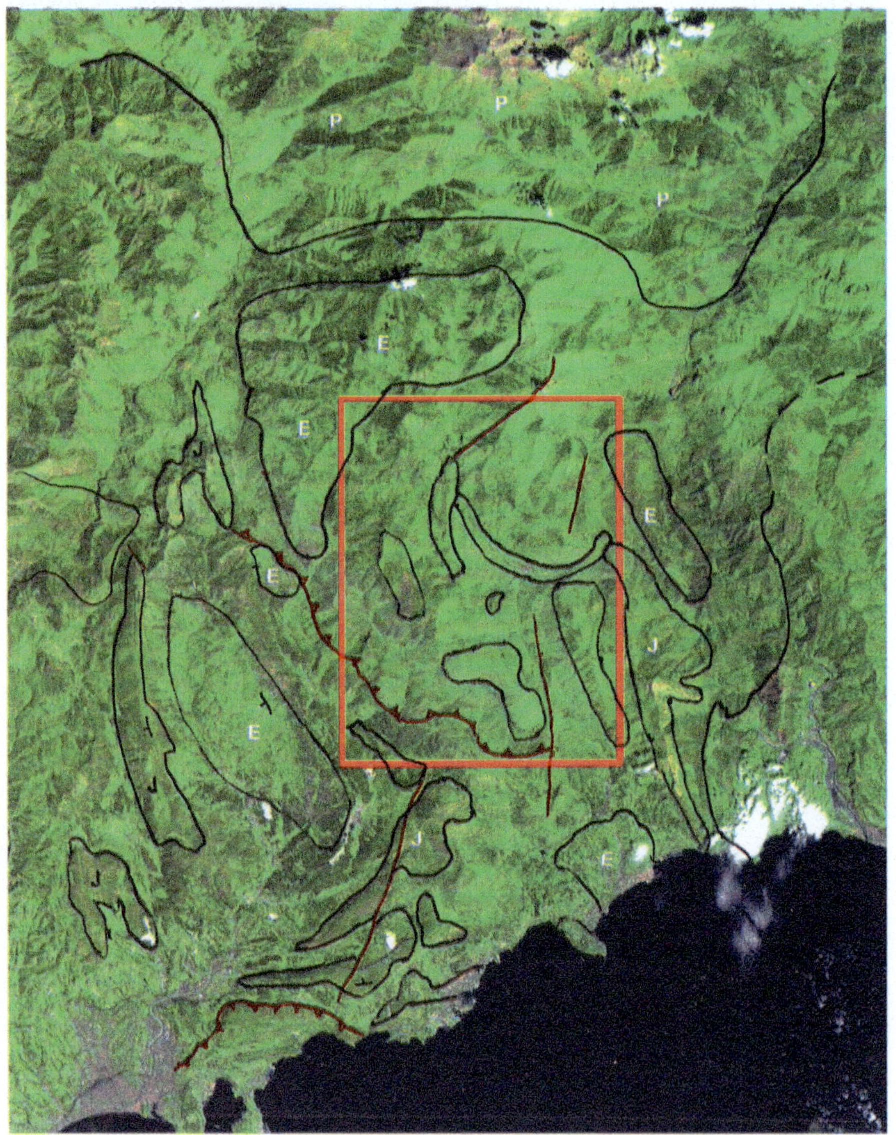

Landsat Image L7 SLC-on, 24 June 2002

Regional Environment

The image covers 800 km^2 in the French Alpes Maritimes on the Italian border north of the peninsula of Menton. The main rock units surrounding the Example site have been delineated using the same codes as in the air photo. The rocks delineated as **P** along the north edge of the image area are Upper Paleozoic sediments that adjoin the Hercynian Mercantour Massif.

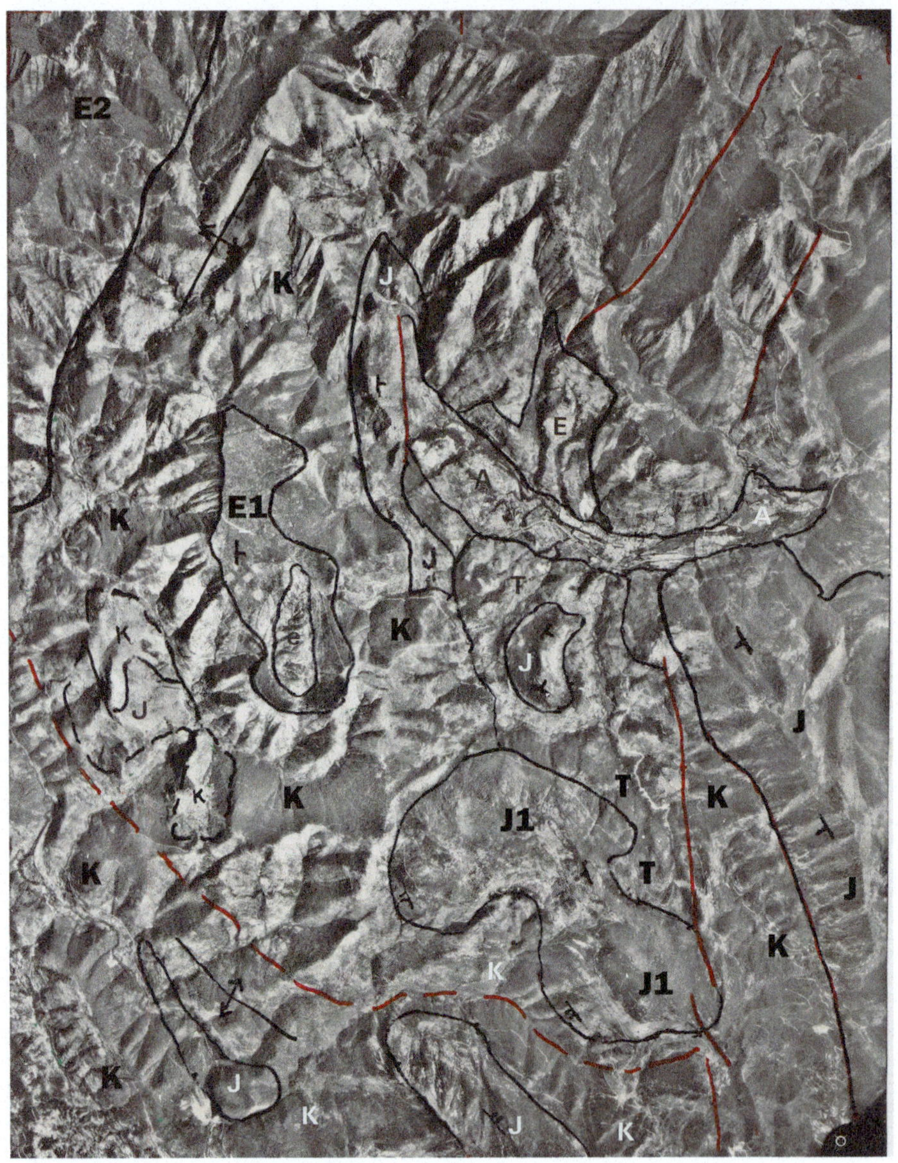

B11 Airphoto

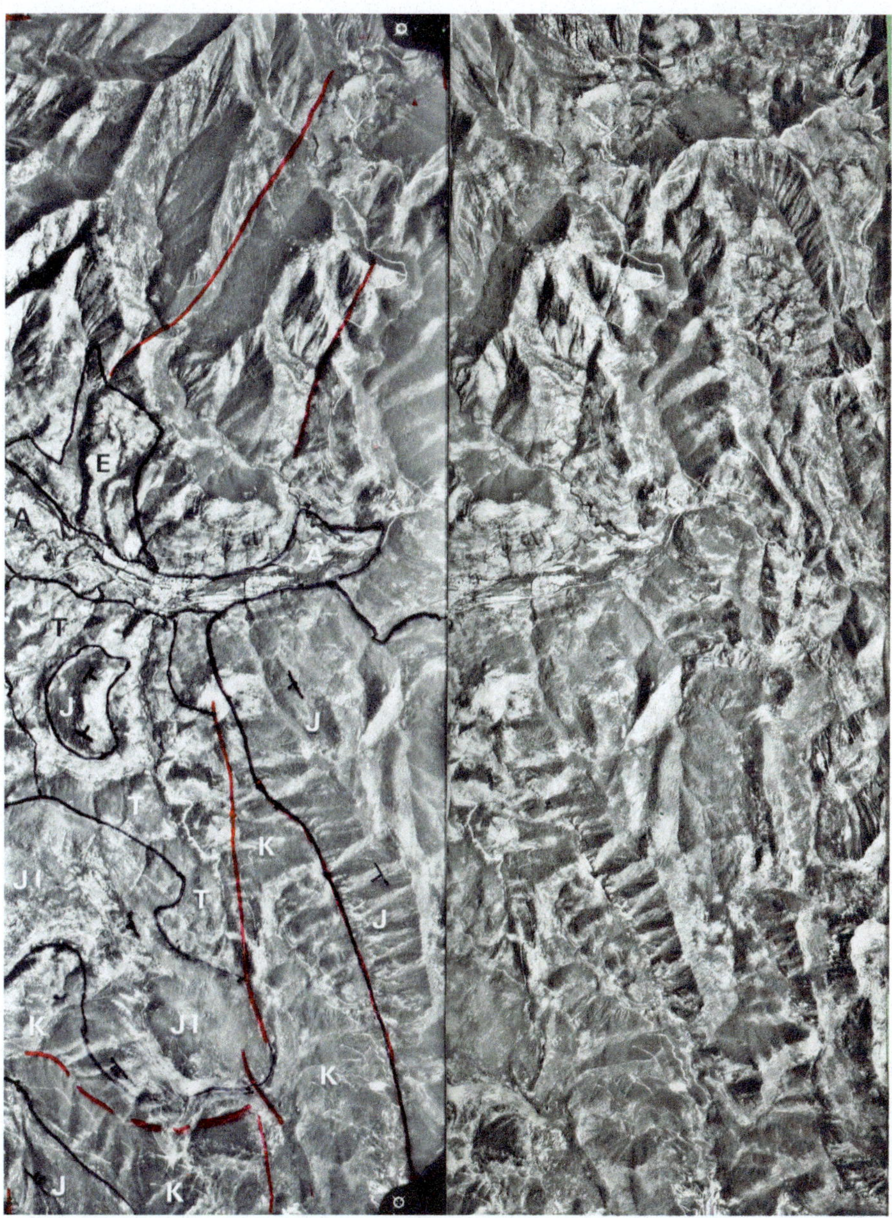

B11 Stereopair

Photo Interpretation

The complexity of this Example lies in the general erodibility of the occurring sedimentary rocks.

Occurrences of five Mesozoic and Tertiary sedimentary rock units are delineated in the 135 km^2 area of the 1:35,000 scale air photo. They range from 500 m to 1,000 m elevation from south to north.

They have been variously folded and thrusted by the Mid-Tertiary Alpine Orogeny. All post-Triassic rocks in the photo area have been displaced southward by the main Alpine events including some gravity sliding over the plastic Triassic sediments.

An east-west Late Tertiary compression related to the Pyrenean Orogeny produced some thrusting and uplifting of the resistant Jurassic limestones. With the exception of the Jurassic limestones all of the units are strongly eroded due to their relatively weak composition. There is little to no morphologic evidence of folding—two anticlinal axes have been transposed from the geologic map.

K—These are the country rocks of the regional *Arc de Nice*. They are relatively weak Upper Cretaceous marly limestones. Their intense erosion is due to their poor consolidation.

T—The oldest sediments in the area are Upper Triassic marl and gypsum.

J—There are six occurrences of thick, less dissected Upper Jurassic Portland limestone. **J1** is the most prominent occurrence; it stands among its surroundings as an extrusive-appearing mass. Its elevations are 1,180 m at the north end and 1,240 m at the south. Adjacent terrains range from 500 to 950 m.

E1 & E2—These-Eocene compact limestones and Oligocene flysch and sandstones occur in relief inversion on the Cretaceous beds.

A—These are recent alluvial deposits of the Bevera River at Sospel.

Photo Source

Copyright IGN, 1979, 7021, 68-69

Glacial and Alpine-Glacial Terrains

This category of complex terrains contains 15 examples. Glacial types are represented by seven examples in Canada; two in the Arctic Islands, four in the Shield and one in the Maritime Appalachians.

The Alpine-glacial set contains one example in the French Alps, two in the Yukon Cordillera and five in the Torngat Mountains of Labrador.

The definition and descriptions of glacial landforms mentioned in this category can be found in Prest V. K. (1983) Canada's Heritage of Glacial Features, Geological Survey of Canada Miscellaneous Report 28.

Glacial Terrains

All these examples contain glacial and glaciofluvial deposits of the Late Pleistocene (Wisconsin/Würm) glaciation and site-associated occurrences of Holocene, alluvial, lacustrine, marine and periglacial deposits.

Figure G1: North Baffin

Location 72°03′N, 80°23′W

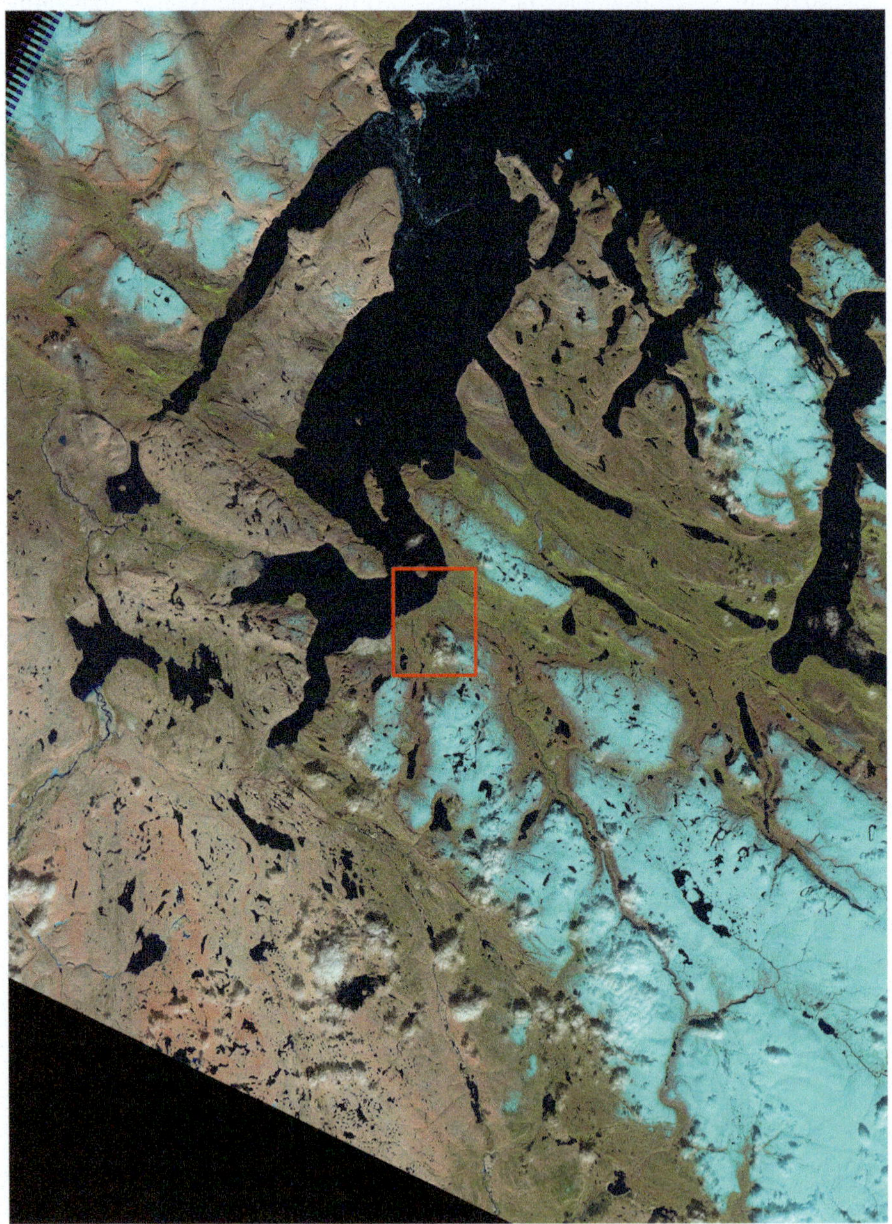

Landsat Image T/M 4-5, 16 August 2003

Regional Environment

The North Baffin area is in the northeast Canadian Shield. The area is at Koluktoo Bay at the head of Milne Inlet fjord.

Structure

The image area is divided into three major structural units. Archean granitic gneisses occur in the snow-covered southeast. The terrain to the north is the Baffin Rift Zone. Proterozoic sedimentary rocks are block-faulted there into northwest striking horst and graben lakes and inlets. The rocks on the south are Ordovician limestones of the Phillips Creek Trough seen on Figures G2 and Zi4 40 km southeastward.

Physiography

The Archean area in which the example site is located is at the snow-covered north end of the Baffin Upland on the south edge of the rift zone. It has generally low relief and rises in elevation to 700–800 m. The Rift zone to the north is at 150–300 m in elevation excepting the snow-covered area which is an outlier of Archean rocks that rise to 900 m.

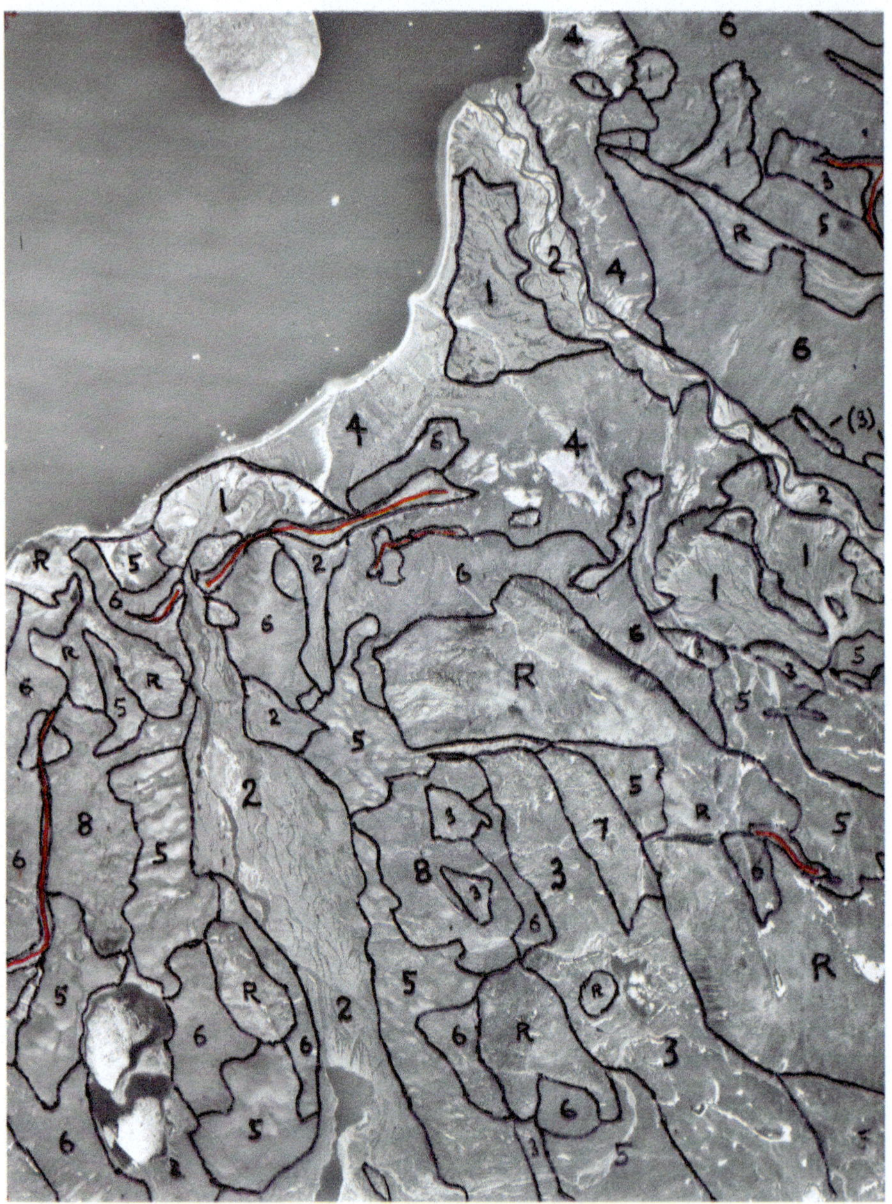

G1 Airphoto

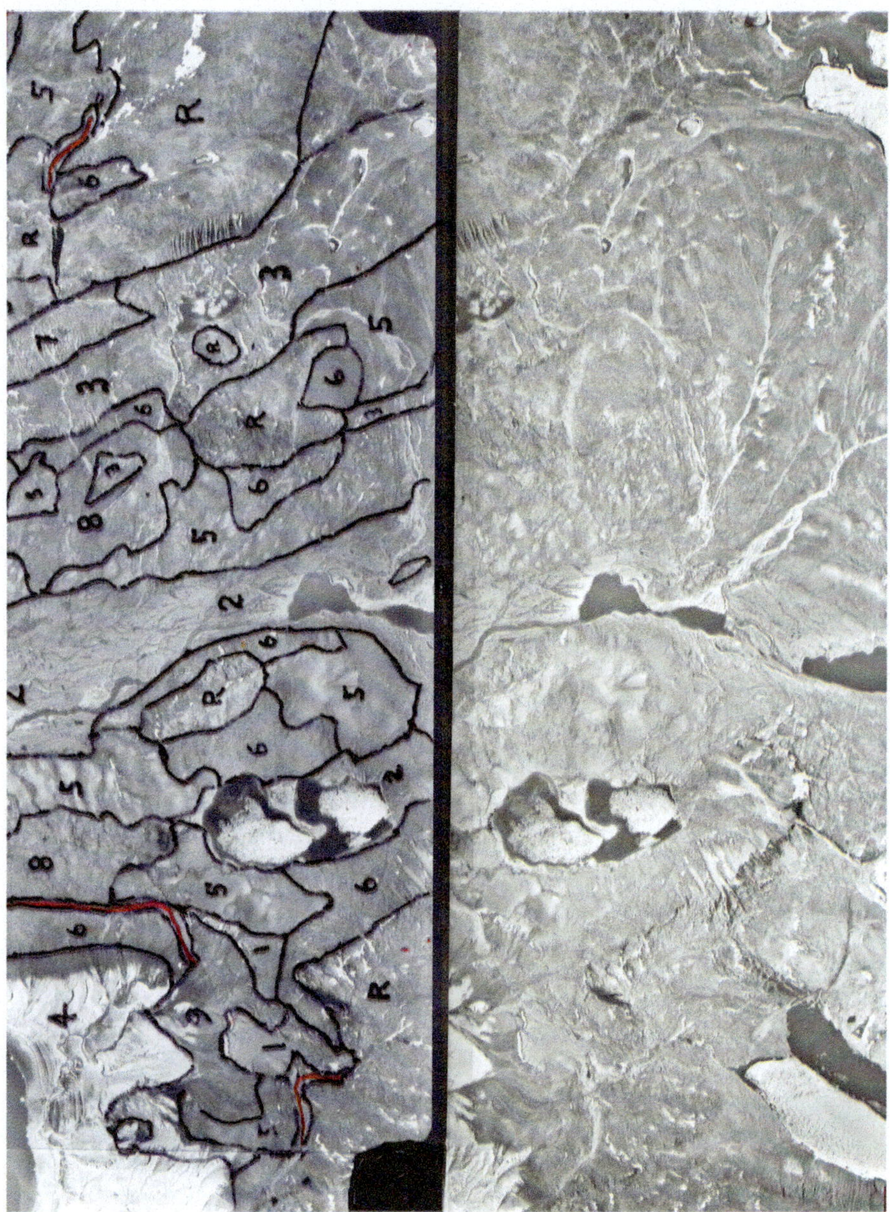

G1 Stereopair

Photo Interpretation

This 1:60,000 scale photo area covers 145 km^2 and is characterized by features of deglaciation and periglaciation. Twenty-nine terrain units of five genetic groups have been delineated.

Bedrock

5—There are five occurrences of Archean outcrops. The largest, in the center and lower right, rises from 440 to 700 m elevation.

Marine Deposits

4—The main occurrence of these sand and gravel deposits rises from a 13 m high bluff of raised beaches at the shoreline to 60 m elevation inland. The unit to the north is at 40 m elevation.

Glacial Deposits

5—ablation moraine Two units of this moraine occur on the west and south flanks of the bedrock hills; two other units are on the west side of the large outwash plain. The material is non-stratified and contains a heterogeneous mixture of particle sizes.

8—re-worked ground moraine This unit has been modified by meltwater scour. It occurs west of the ablation moraine deposit beside the outwash plain.

Post-glacial Deposits

1—raised deltas Five occurrences of these bedded sand and gravel deltas are unevenly distributed in the photo area. Two are at the north and west end of the marine deposits. The north unit is at 15 m elevation; the west unit is higher at 45 m. The eastern units are at two levels at the inland contact with the marine plain and at the outlet of an off-photo outwash plain. The higher unit lies 45 m above the marine plain, the surface rising from 120 to 195 m elevation. The adjacent unit is lower at 70–150 m. The last unit, in the northeast corner of the photo, is at 145 m elevation.

2—outwash plain Two units of this plain occur. The larger contains well sorted braided sands and gravels, and is 0.75–1.5 km wide. The narrower meandering type northern unit is 400–800 m wide.

3—kame terrace This large unit of stratified sand and gravel which is nearly 6 km long and 1.5 km broad developed between the margin of a Wisconsinan glacier and the bedrock hills.

Eskers—A red line traces narrow elongate ridges of subglacial meltwater sand and gravel.

Periglacial Deposits

6—gelifluction slopes Ten occurrences of this process are scattered over the photo area. They are described as Unit **Zm1.1** in the Periglacial category.

7—ice wedge polygons There is a single occurrence of this type of periglacial form, located between the kame terrace and the large bedrock area. These polygons are described as Unit **Zi4** in the Periglacial category.

Photo Source

Courtesy of National Air Photo Library, A16216, 145-146

Reference

Klassen, R. A. (1993). Quaternary geology and glacial history of Bylot Island, Northwest Territories. Geological Survey of Canada Memoir 429.

Figure G2: Krag

Location 71°32′N, 80°17′W Northwest Baffin Island Canada

Landsat Image T/M-5, 16 August 2003

Regional Environment

The area is 55 km south of Figure G1.

Structure

The site lies 30 km south of Figure G1 in the Phillips Creek Fault Zone. The bright zone on the south are Ordovician limestones of the Phillips Creek Trough, a 16,000 km^2 fault bounded block. The pink 600–800 m Archean Krag Mountains separate the Phillips Creek area from the structures of the Rift Zone of Figure G1. The site is 40 km northwest of Figure Zi4.

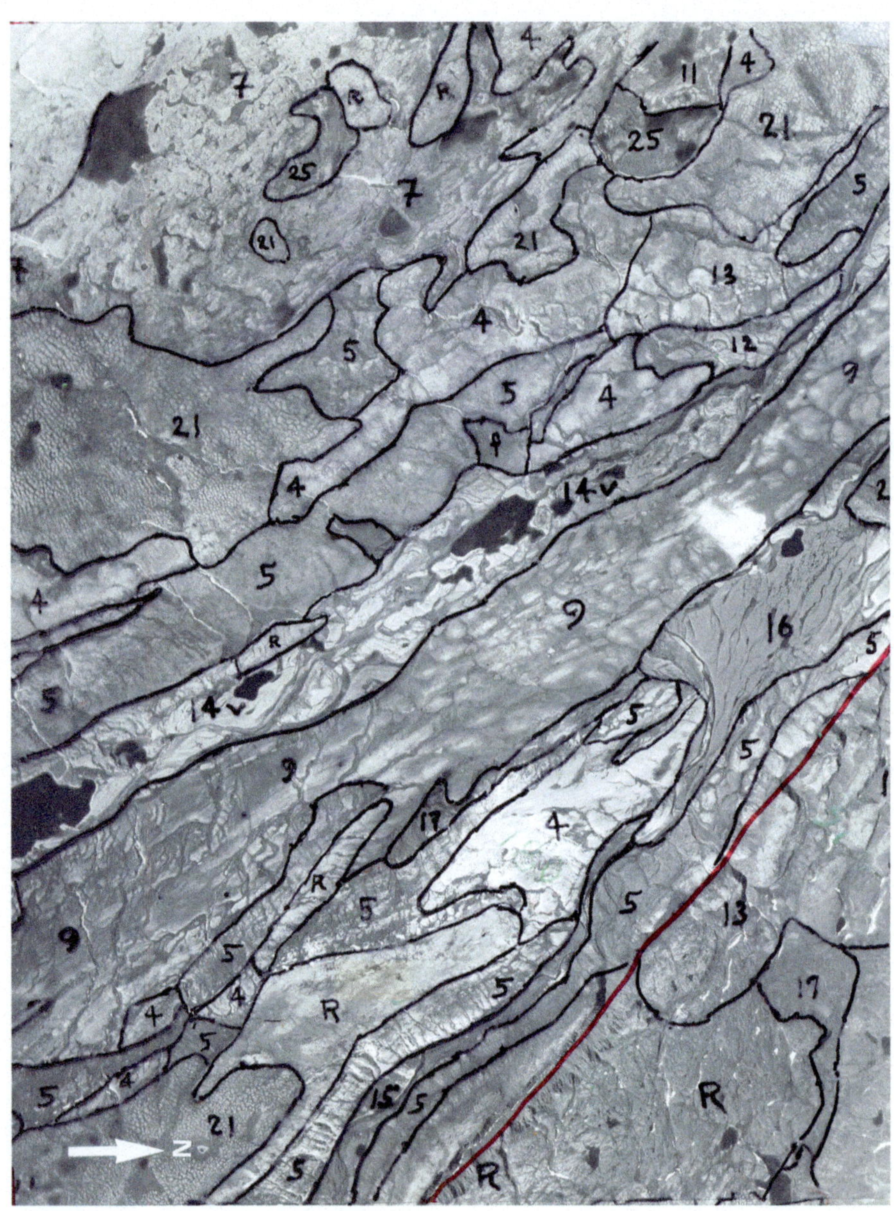

4
7
R
R
11
4
25
21
25
7
21
21
5
13
5
4
12
5
4
9
21
4
14v
5
4
9
16
R
5
14v
5
5
9
17
4
R
5
5
13
9
5
4
17
R
4
5
5
4
5
21
15
5
R
R
5
N

G2 Airphoto

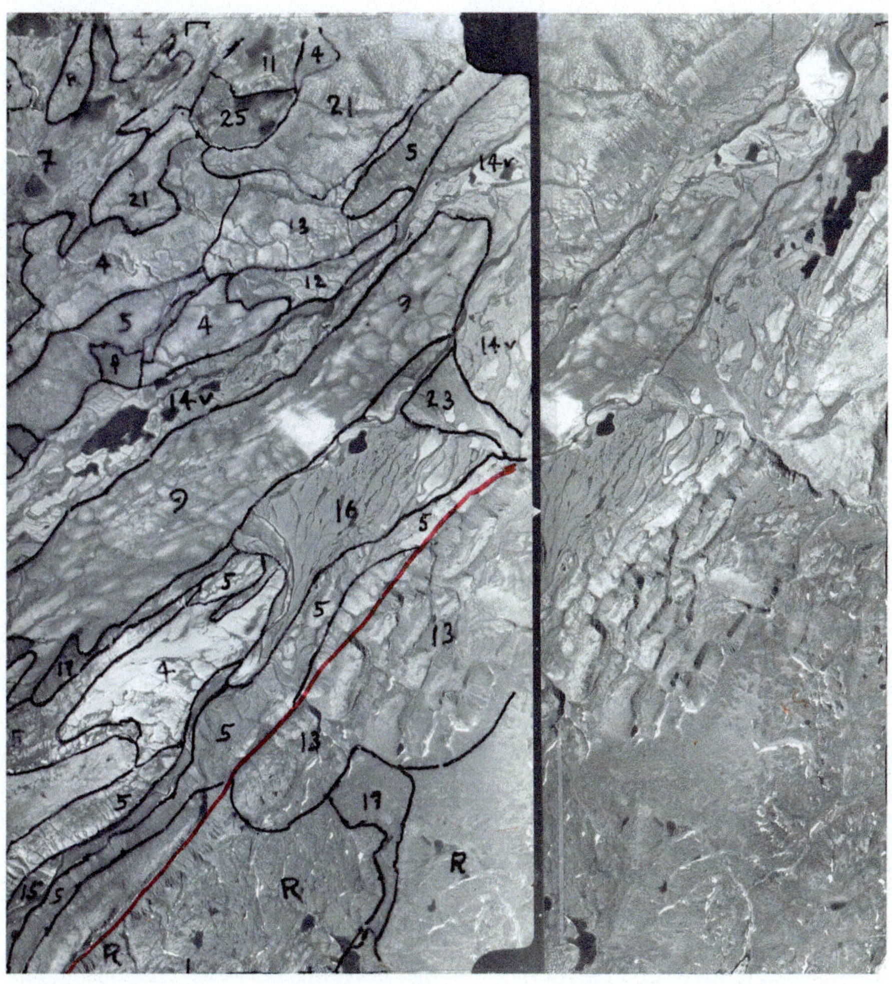

G2 Stereopair

Photo Interpretation

The 1:60,000 scale photo area covers 145 km^2. The complex site includes four units of glacial deposits, five units of glaciofluvial deposits, a unit of periglacial ice wedge polygons and two bedrock units.

Glacial Units

4—**Ground moraine** occurs on the limestone plateau at 290 m elevation and on the margins of the valley.

5—**Reworked moraine** occurs along the side of the main valley and a narrow valley that lies along the edge of the Krag highland.

7—is **hummocky moraine** on the plateau surface.

9—a long broad zone of **drumlins** lies in the valley at 115 m elevation.

Glaciofluvial Deposits

11—is a small **kame** deposit associated with a lacustrine depression at the west edge of the photo.

13—Are two units of **ice marginal meltwater channels.** The larger is etched into the rocks of the side of the Krag upland, while the smaller occurrence is cut into the low scarp of the limestone plateau.

14v—**outwash plain** of glacial meltwater sands and gravels are deposited down the floor of the fault valley at 115 m elevation (visible in the stereo view).

15—**outwash terraces** are in a narrow valley that flows along the margin of the regional fault

16—This **delta** of the terraced stream 15 is at elevation 105 m. It occurs at a point where the stream deposits were blocked by the alluvial fan, unit **23**, of a stream that drains a 3 km^2 lake at 300 m elevation on the gneiss upland 3 km to northeast off-scene. The stream diverged around the delta mass to flow along the base of the 10 m higher drumlin belt.

Periglacial Unit

21—Two extensive occurrences of **ice wedge polygons (Zi4)** are developed on the glacial tills of units **4** and **7** at 290 m elevation on the edge of the Phillips Creek unit.

Bedrock

R—This unit, Archean gneiss of the Krag Mountains, is at elevation 395 m.

Fault—The red line traces the normal fault that bounds the northeast side of the valley and is the scarp of the 300 m high Krag Mountains

Photo Source

Courtesy of National Air Photo Library, A16216, 139-140

Reference

Rand Corporation (1963) A report of the physical environment of northern Baffin Island and adjacent areas, Northwest Territories, Canada. Memorandum RM-2706 -1- PR.

Figure G3: Natkusiak

Location–72°51′N, 110°06′W

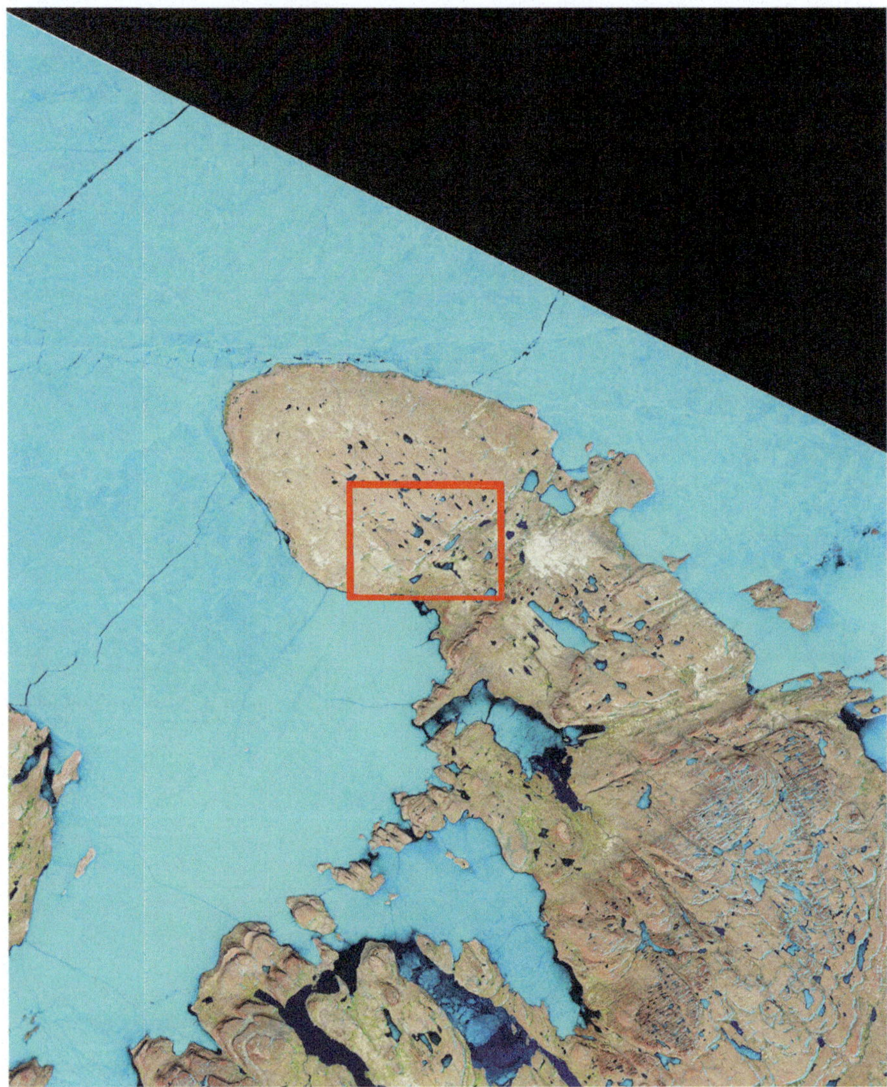

Landsat Image T/M 4-5, 17 July 2011

Regional Environment

The scene covers a peninsula on the north coast of Victoria Island of the Northern Canadian Plains.

Structure

The site area is on Natkusiak Peninsula at the eastern end of a cross-island 30–50 m high synclinal belt, the Shaler Mountains. They consist of Proterozoic sedimentary rock interbedded with gabbro sills. The fault crossing the air photo is the north limit of the mountains which cover the rest of the scene. The remainder of the island is an interbedded sequence 150–300 m elevation of Cambrian and Silurian sedimentary rocks generally covered by glacial deposits.

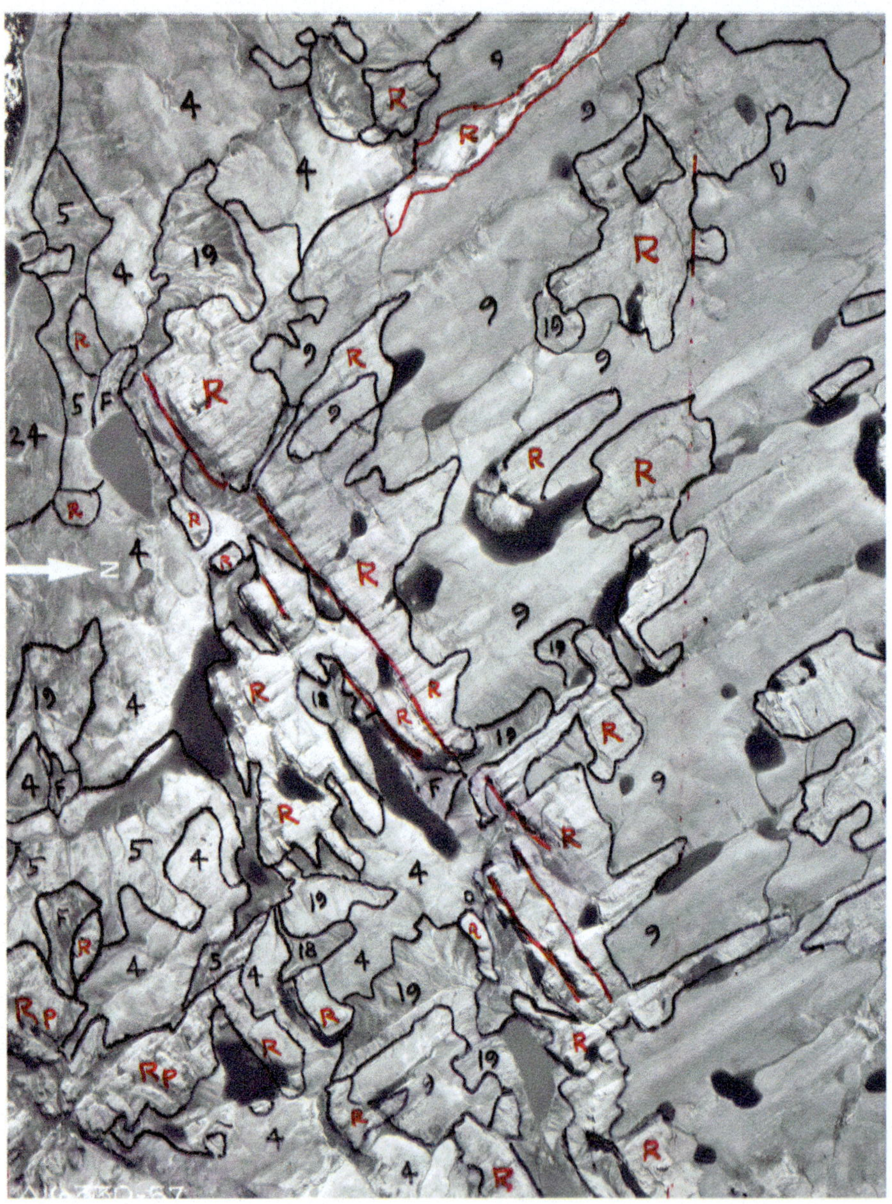

G3 Airphoto

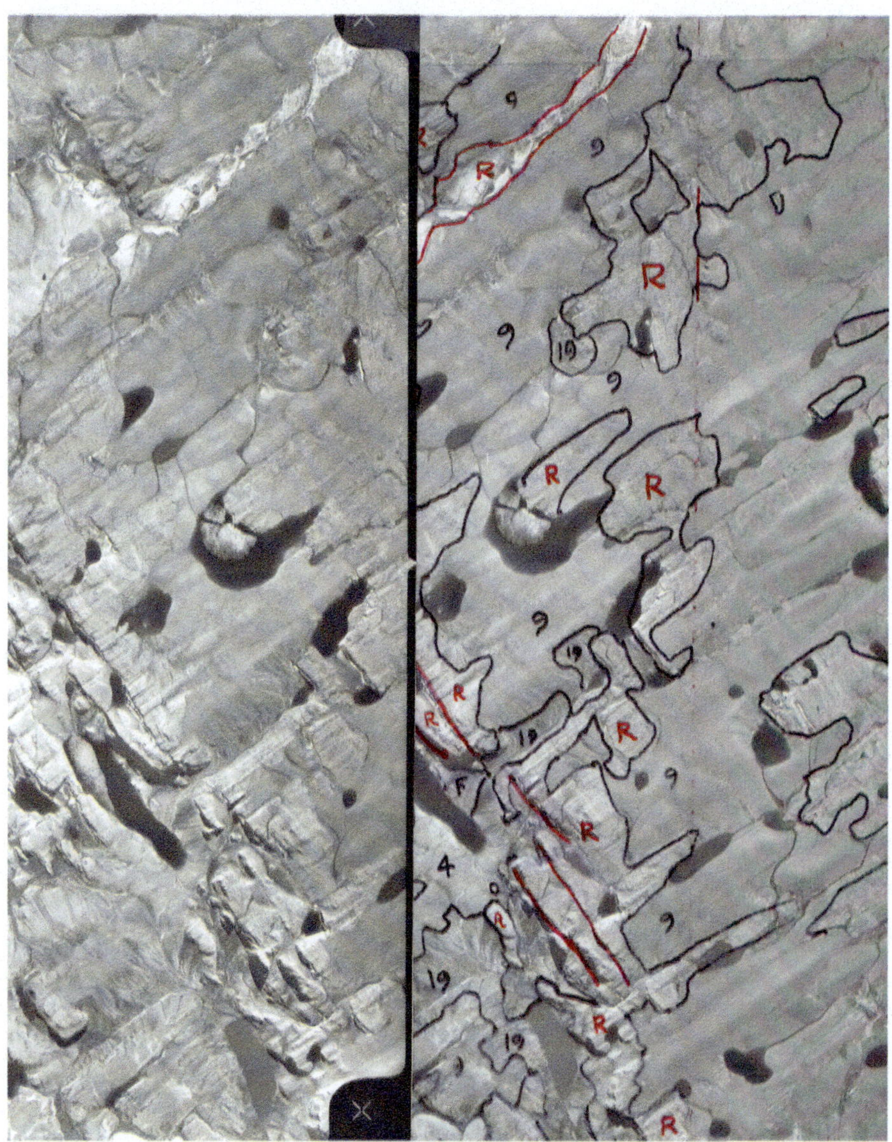

G3 Stereopair

Photo Interpretation

This 1:50,000 scale photo covers 88 km^2. The north coast of Wynniatt Bay, covered with residual ice on 11 August 1958 is at the south edge of the photo.

Three glacial, one periglacial and two bedrock units occur in the photo area.

Glacial Deposits

4—Large deposits of featureless **ground moraine** occur near the bay and on the southeast part of the photo. Deposits are generally thin, ranging from 20 to 100 m in elevation according to local bedrock relief.

5—Only two occurrences of **reworked moraine** are mapped in the south half of the area.

9—The north part of the photo area is dominated by **drumlins** at 115–160 m elevation in crag-and-tail form on the down-ice side of scoured rock ledges, oriented by southeast to northwest glacial flow

19—There are scattered occurrences of Periglacial **gelifluction stripes (Zm1.2)** on the low slopes of till deposits.

Bedrock Types

R—These are outcrops of the Lower Paleozoic sedimentary rocks at 120–140 m elevation.

Rp—These are two outcrops of the Shaler Mountains Proterozoic rocks. The **red lines** mark the possibly fault contact zone of the two rock types

Photo Source

Courtesy of National Air Photo Library, A16330, 57-58

References

Fyles, J. D. (1963). Surficial Geology of Victoria and Stefansson Islands, District of Franklin. *Geological Survey of Canada Bulletin 101*.

Rand Corporation. (1963). A report of physiographic conditions of Eastern Victoria Island and adjacent areas, Northwest Territories Canada. Memorandum RM-2707-1- PR.

Figure G4: Rivière Jacob

Location 47°46′N, 70°12′W

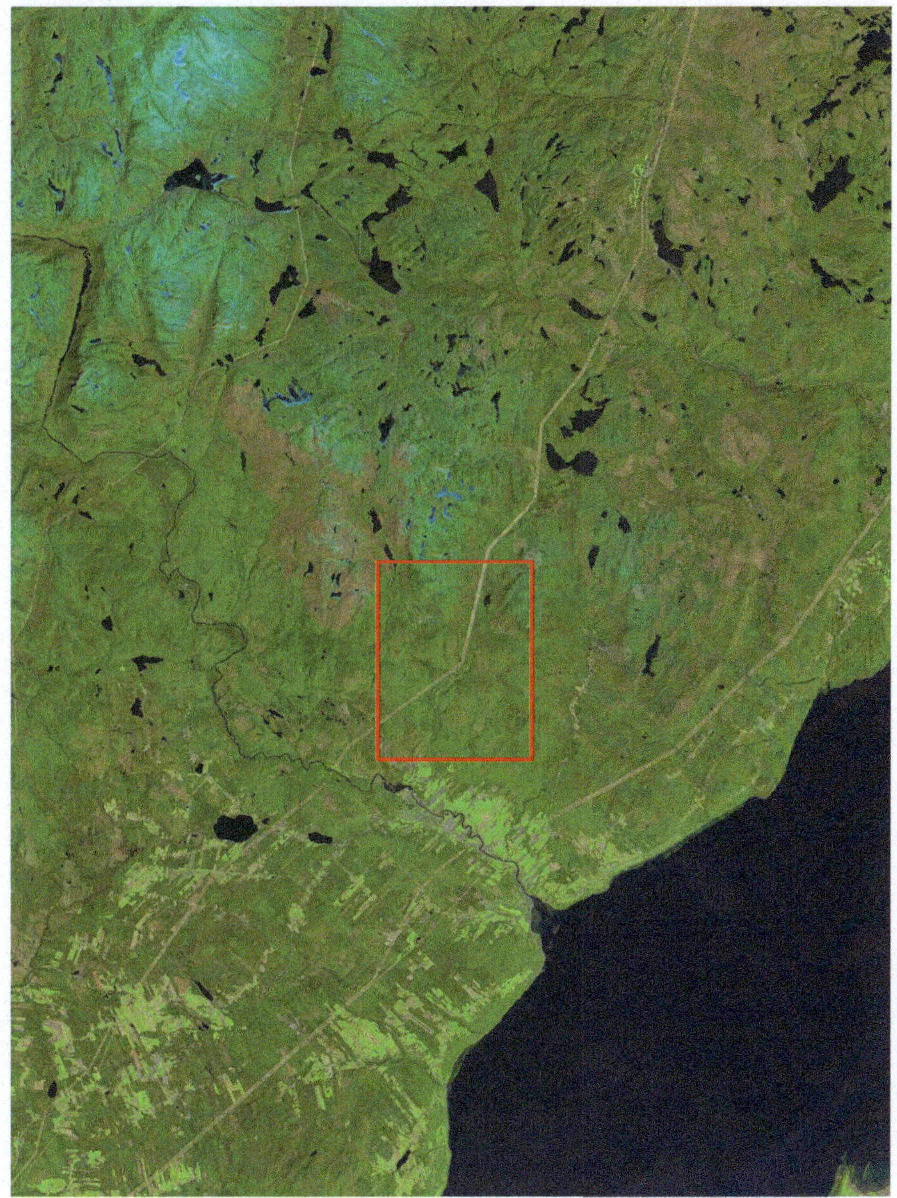

Landsat Image T/M 4-5, 17 May, 2010

Regional Environment

Structure

The example area is located in Proterozoic gneiss terrain of the southeast Shield Grenville Province.

Physiography

The region is in the low mountainous terrain of the Laurentian Highlands, 20 km north of the Charlevoix Impact Crater on the St Lawrence River. The north part of the crater is not resolved in the scene. Rock hills range from 300 to 700 m in elevation, with an outwash valley at 400 m. The green areas are spruce forest.

The 200 m wide white band traversing the area is a hydro-electric power transmission line.

14v
5
R
R
6
5
5
6
R
5
6
17
5
14v
R
R
6
4
6
R
17
6
4
17
6
R
R
R
R
14v
17
R
6
4
R
R
6
R
6
R
17
11
4
4
5
R
R
17
4
17
R
6

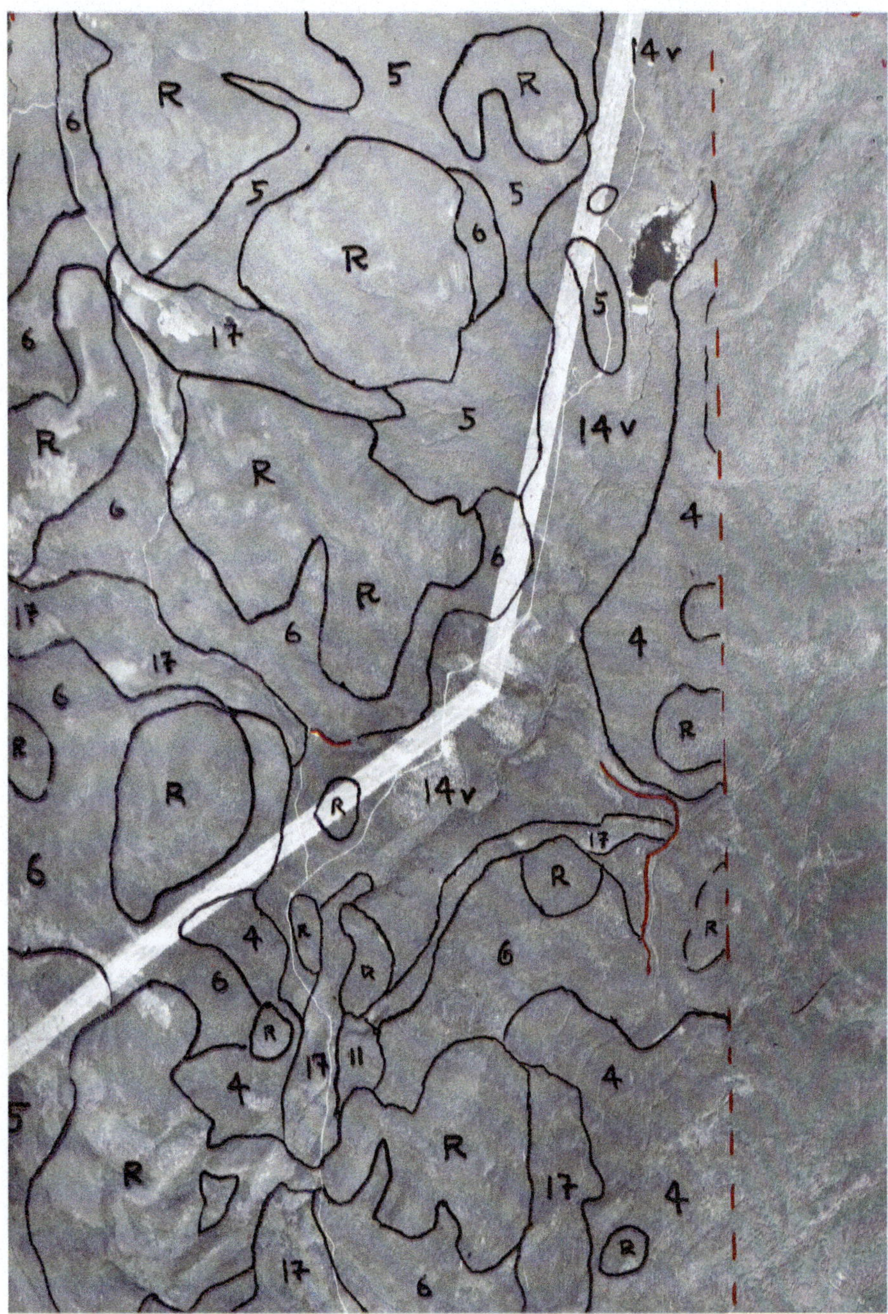

G4 Airphoto

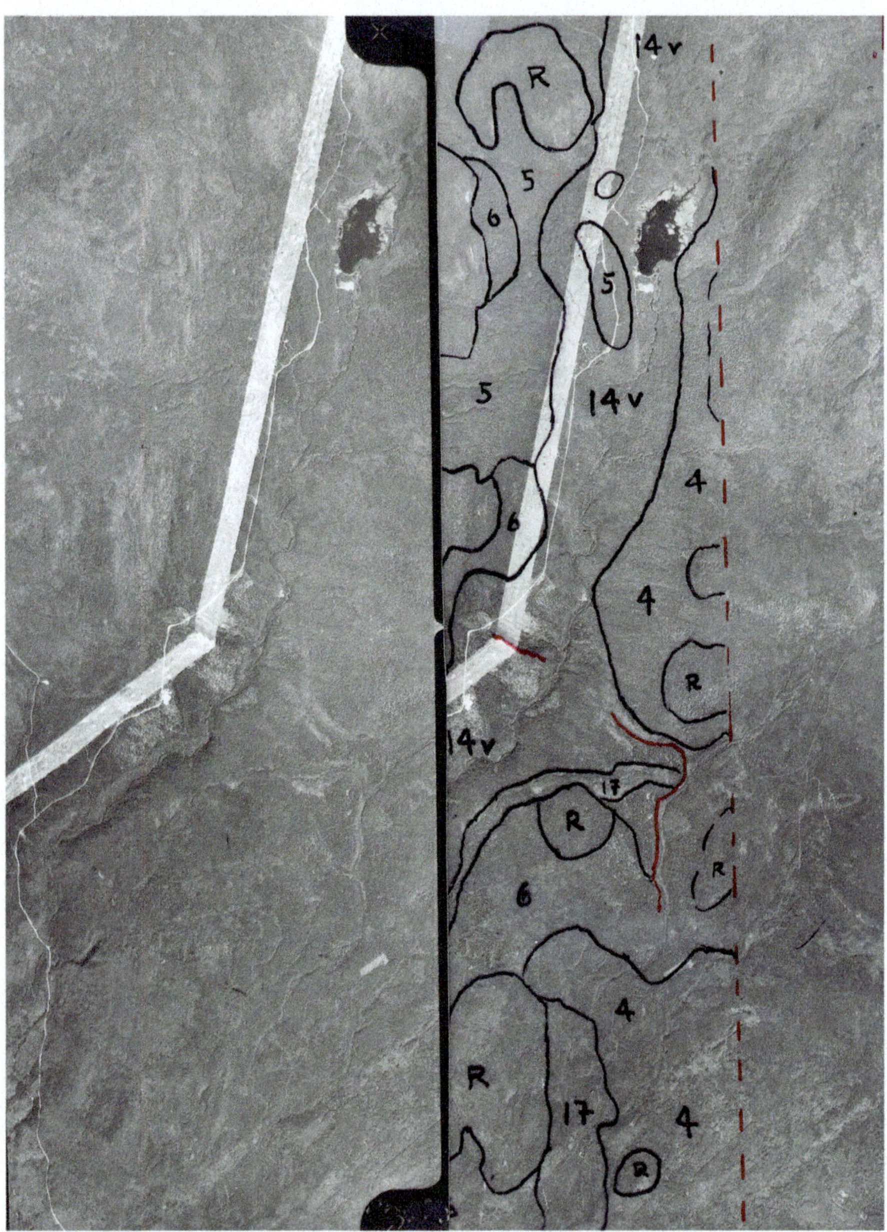

G4 Stereopair

Photo Interpretation

The 1:40,000 scale photo area covers 56 km^2. Three glacial and four glaciofluvial units occur in the site.

R—denotes bedrock outcrops.

Glacial Deposits

4—There are only three areas of typical **ground moraine**.

5—Reworked moraine terrain is more common, associated with Unit **17** spillway channels.

6—Veneer moraine occurrences are located adjacent to bedrock hills.

Glaciofluvial Deposits

10—A **red line** traces a 1.5 km long **esker** on the east side of the site.

11—A single **kame** is adjacent to a spillway channel.

14v—This is the main unit; an **outwash valley train**, an 800–1,000 m wide valley that crosses the entire site.

17—There are three **spillway channels** that lead to the outwash valley.

Photo Source

Personal archive

Figure G5: Groswater Bay

Location 54°26′N, 58°16′W

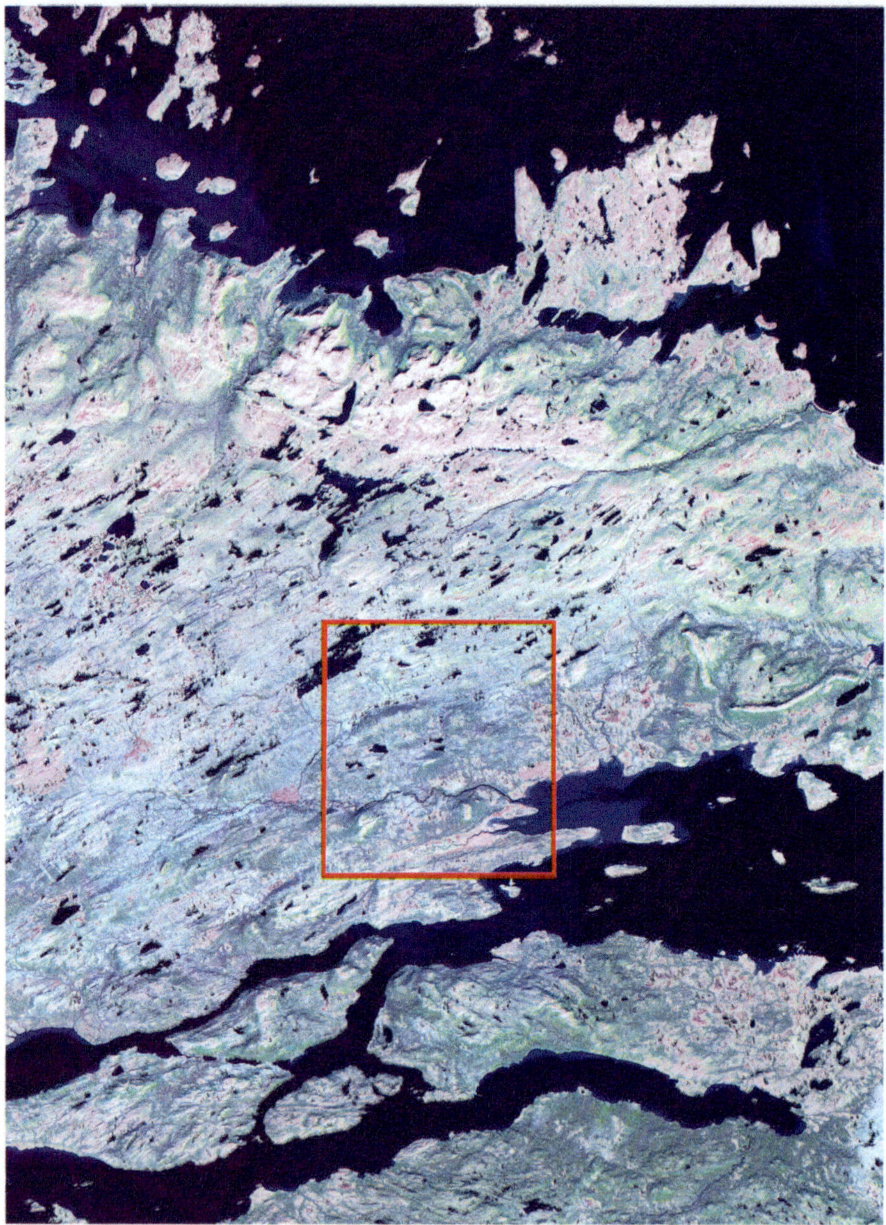

Landsat Image T/M 4-5, 13 September 1987

Regional Environment

The Example area is near the eastern end of the Shield Grenville Province.

Structure

The site is in the arcuate Mid Proterozoic Groswater Terrane. The terrane lies between the Trans-Labrador batholith on the north and the Melville Terrane with Groswater Bay on the south. Elevations range from sea level to 275 m. The southern end of the Labrador Sea is at the north.

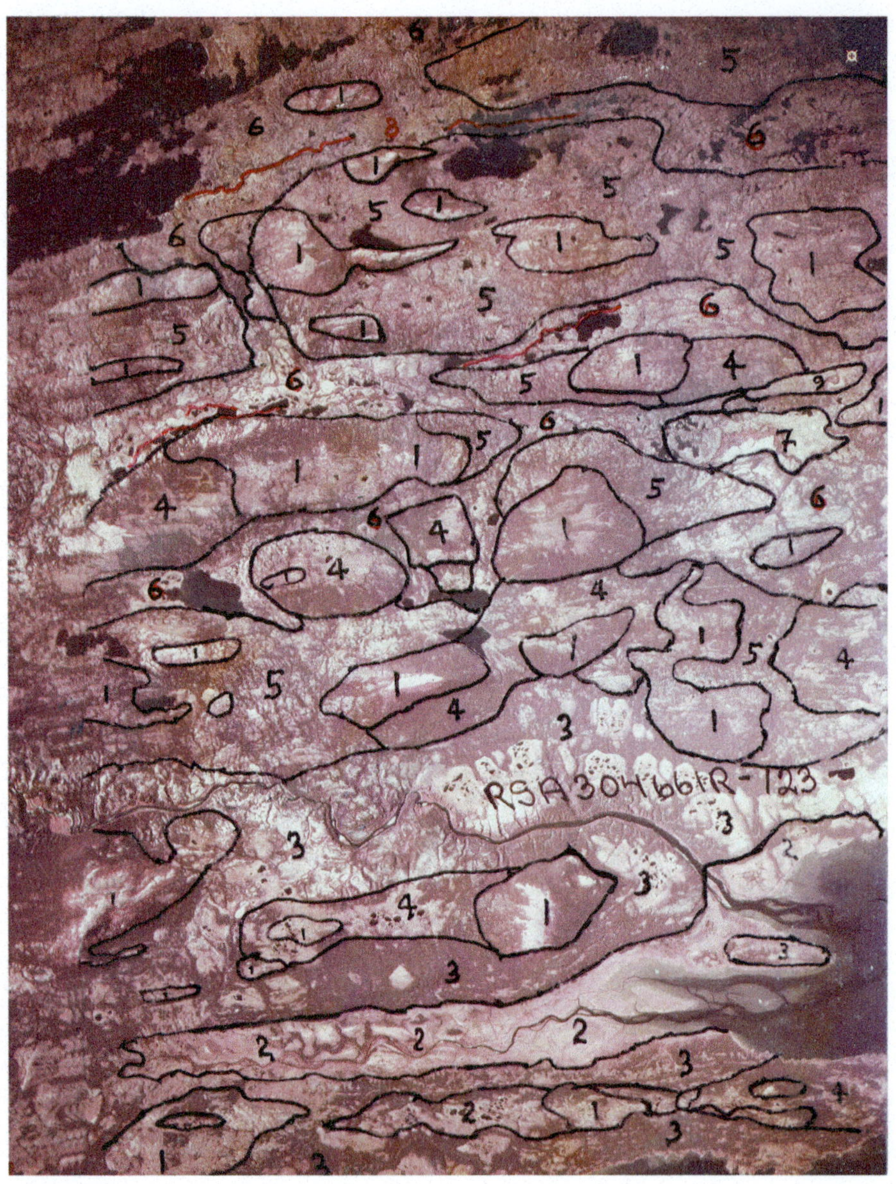

G5 Airphoto

G5 Stereopair

Photo Interpretation

The 1:120,000 scale photo area covers 540 km^2. Eight terrain units, generally aligned according to the east-northeast trend of regional ice movement occur in the site area: Two glacial, three glaciofluvial, two glaciomarine units and one bedrock type (code 1 on air photo). The dark brown areas are forests of black spruce and balsam.

Isolated gneissic **bedrock** hills 200–275 m elevation are scattered throughout the site area.

Glacial Deposits

4—There are five occurrences of **ground moraine** across the center of the site.

5—Extensive deposits of **hummocky moraine** are distributed across the center and north of the site.

Glaciofluvial Deposits

6—A branching network of elongate and broad erosional **meltwater channels** and associated eskers are spread across the center and north parts of the site.

7—There is a single deposit of **outwash** sand and gravel lying in a meltwater channel near the east edge of the site.

8—**Esker** ridges traced in red are located in the meltwater channels.

Glaciomarine Deposits

3—**Glaciomarine sediments** silts clays and associated wetlands occur along the south part of the site. They range from sea level to 130 m elevation inland at the head of Groswater Bay.

2—A **tidal flat** with a 2 m tidal range extends 17 km up a narrow stream at the head of Groswater Bay.

Photo Source

Courtesy of National Air Photo Library, RSA30466IR, 123-124

References

Klassen, R. A., Paradis, S., Bolduc, A. M., & Thomas, R. D. (1992). Glacial landforms and deposits, Labrador Newfoundland and eastern Quebec. *Geological Survey of Canada*, Map 1814A, scale 1:1,000,000.

Lopoukhine, N., Prout, N. A., & Hirvonen, H. E. (1977). Ecological land classification of Labrador. *Ecological Land Classification Series, 4*, Fisheries and Environment Canada.

McLaren, P. (1980). The coastal morphology and sedimentology of labrador: A study of shoreline sensitivity to a potential oil spill. *Geological Survey of Canada Paper*, 79–28.

Figure G6: Freshsteak Lake

Location 54°38′N, W59°56′W

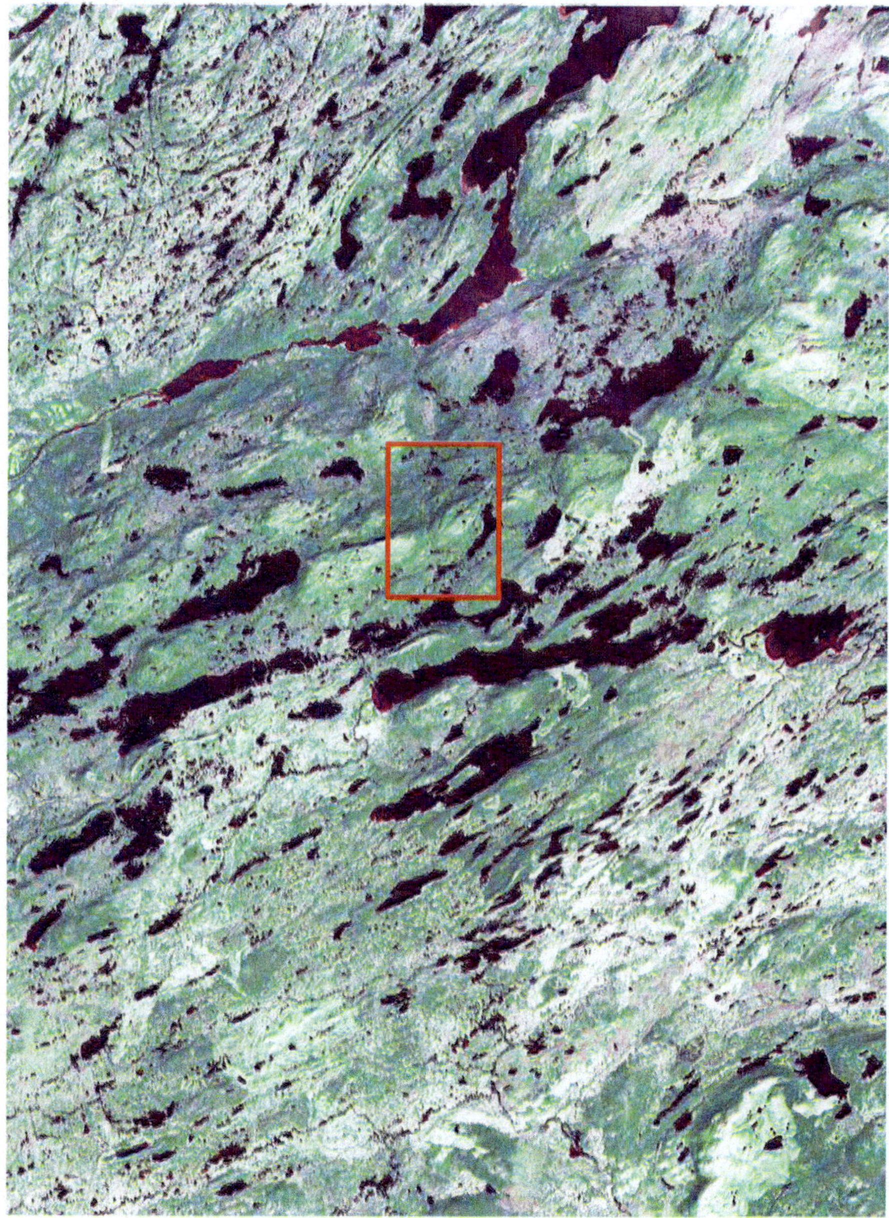

Landsat Image T/M4-5, 20 September 1987

Regional Environment

The Example area is 90 km north of Lake Melville in eastern Labrador.

Structure

The site is in the northern part of the Proterozoic Trans-Labrador Granitoid Batholith, a wedge-shaped crustal segment which extends along the northern flank of the Grenville Province of Figure G5. The Landsat image shows one of a number of northeast striking thrust faults.

Physiography

The site divides into two thrust fault related levels; the area north of the regional strike lakes and ridges is at average 170 m elevation. It is characterized by hummocky and veneer moraine deposits. The area south of the outcrops is at 300 m; general elevation rising to 455 m is interspersed with hummocky, ribbed and veneer moraines.

R
5
6
6
1
R
6
17
6
R
17
7
R
R
R
17
R
6
6
R
7
6
R
R
17
R
6
R
6
7
8
7
R
6
R
17
1
5
7
1
4
6
11
17
5
R
6
R

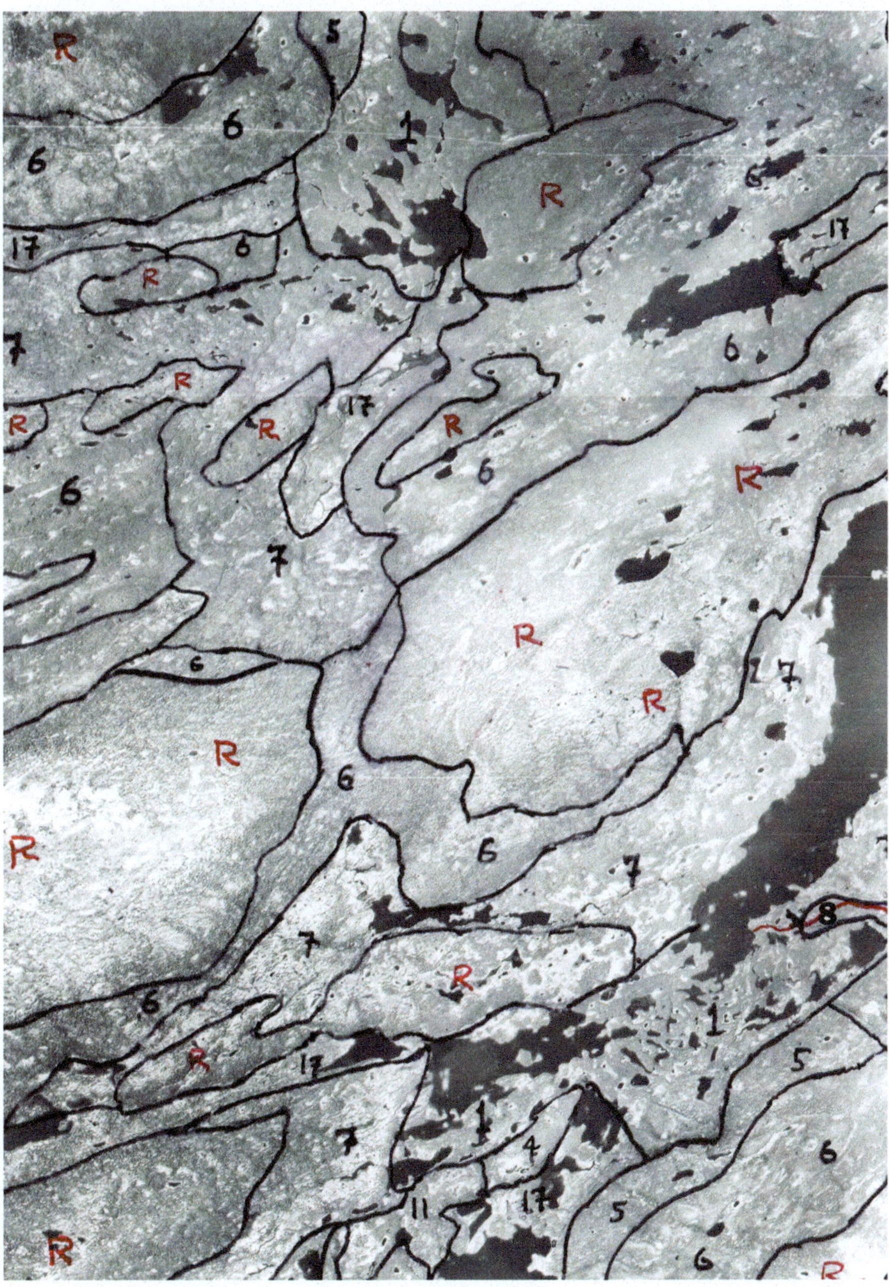

G6 Airphoto

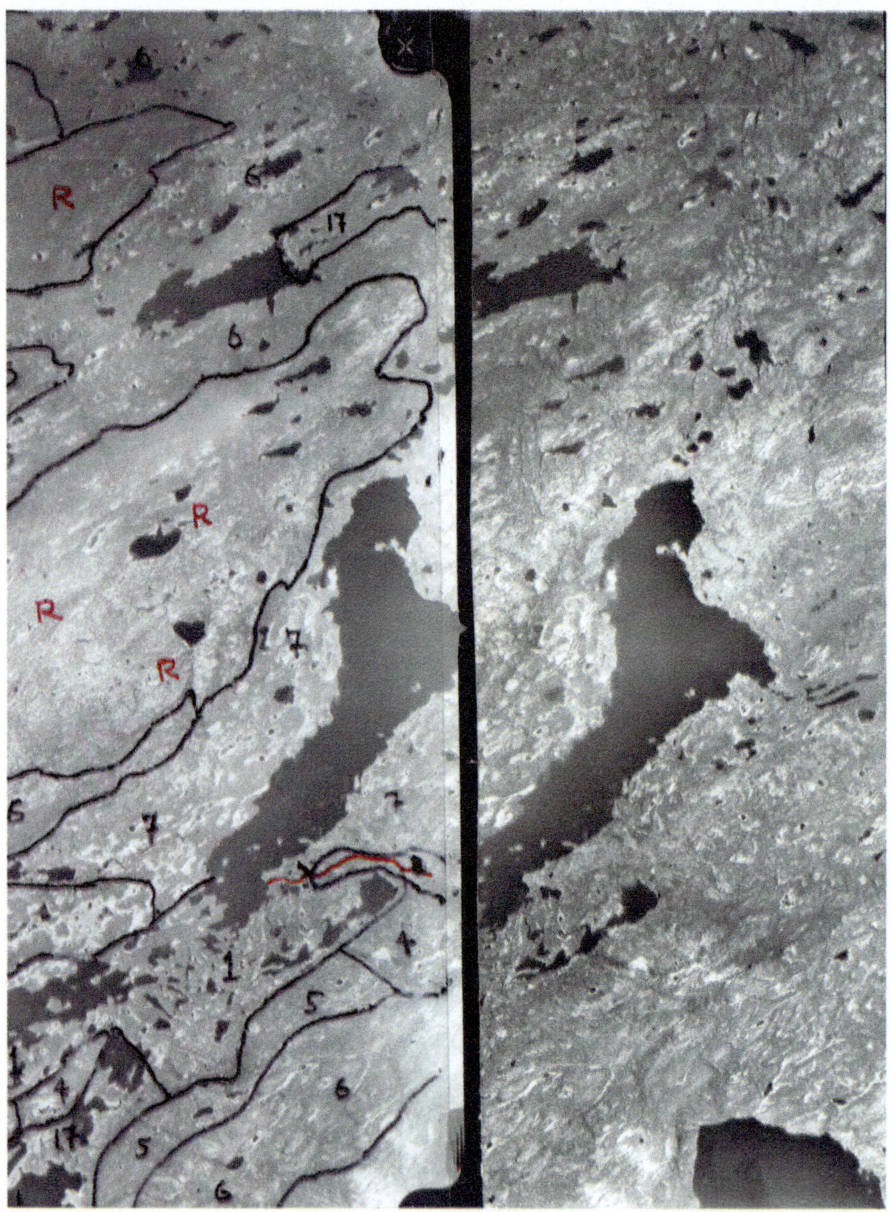

G6 Stereopair

Photo Interpretation

The 1:50,000 scale photo area covers 88 km^2. Ten terrain types occur, including six glacial, three glaciofluvial and one bedrock type.

R—Granite outcrops range from 220 to 455 m in elevation.

Glacial Deposits

1—Two deposits of **ribbed moraine** occur at different elevations. The smaller, in the north, is in the low terrain at 125 m elevation. The larger, 1.5 wide by 4 km long, in the south is at 290 m elevation. Their distinct morphology consists of arcuate ridges separated by finger-like lakes.

4—A single small deposit of ground moraine occurs on the southeast edge of the site.

5—Two occurrences of **reworked moraine** are on the margin of ribbed moraine (**1**) low areas.

6—**Veneer moraine** is extensive in the site. Deposits are associated with and downslope from the bedrock outcrops.

7—**Hummocky moraine** occupies all the terrain below veneer moraine (6) deposits. It is of ablation type, and is non-oriented, devoid of any linear elements. Local relief is typically up to 10 m.

8—A single occurrence of a **marginal moraine** is east of the south end of the lake. It shows 1,200 m of a 3 km long ridge. An esker (**10**) runs along the top of the ridge.

Glaciofluvial Deposits

10—The single **esker** ridge is traced in red along the crest of the marginal moraine.

11—A 1,400 m long by 300–400 m wide 20 m high **kame** ridge occurs at the margin of the spillway channel (17) in lower right of the site area.

17—Six spillway channels of varying lengths leading to ribbed moraine (1) depressions occur throughout the site. The longest is 5.5 km; the others range from 1 to 5 km.

Photo Source

Courtesy of National Air Photo Library, A21898, 110-111

Figure G7: Petitcodiac

Location 45°55′N, W 64°34′W

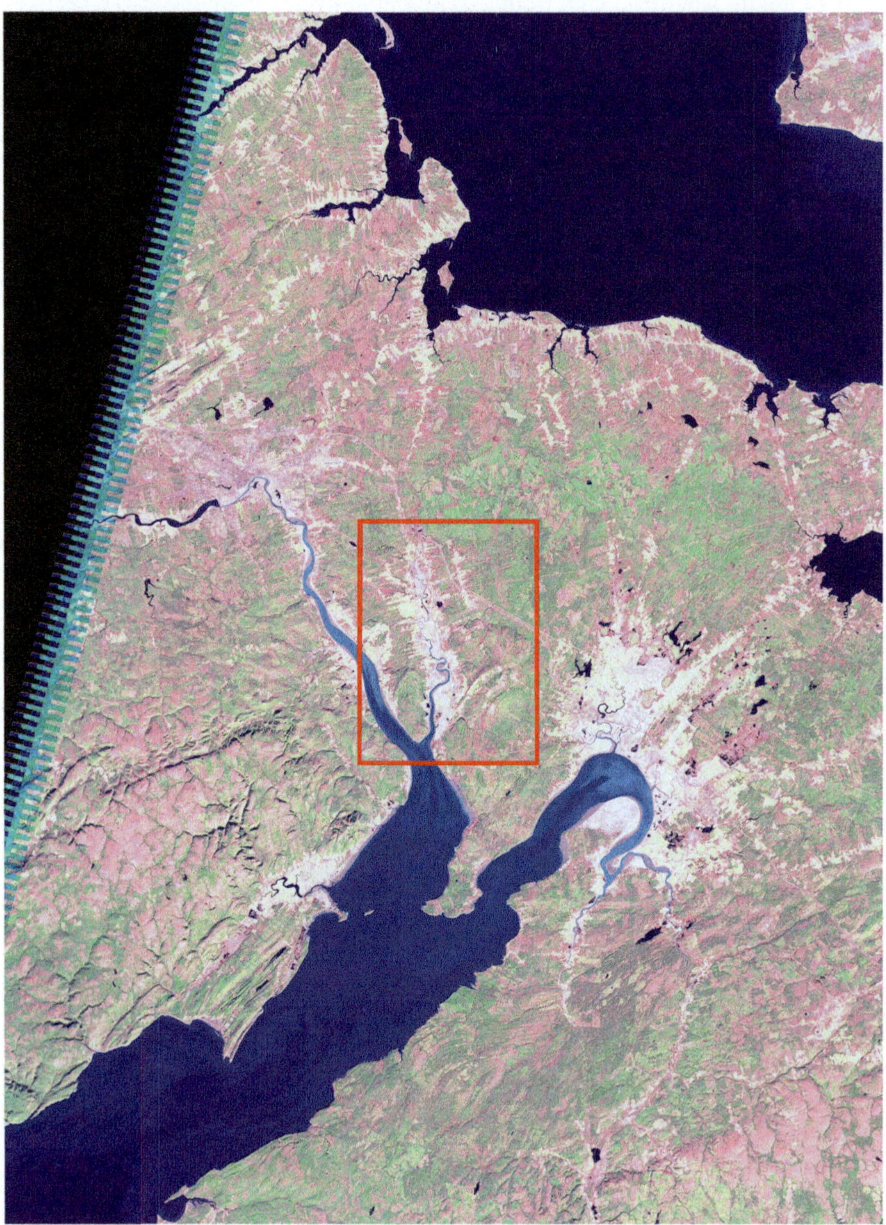

Landsat Image T/M 4-5, 19 November 2009

Regional Environment

The example area is situated at the northern end of the globally highest tides of the 290 km long by 60 km wide Bay of Fundy in southeast New Brunswick. Northumberland Strait is at the north.

Structure

The site is situated in gently deformed Lower Carboniferous clastic rocks of the Maritime Appalachians. The 10–15 m wide tidal Petitcodiac River flows into the Shepody and Chignecto Bays which in turn flow into Fundy Bay's tides. The site contains a limited number of complexly distributed terrain types.

Mv
Mv
Mb
Mb
Ma
Mv
R
Mv
Mb
R
W
Mb
Mb
Mv
Mv
Mb
Mv
Mb
W
R
Mb
Mv
R
R
Mv
Mb
Mv
Mv
Mb
Mb
Mv
Mv
R
W
Mv
R
Mb
Mv
R
Mv
Rl
W
R
Rl
R
R
Mb
Mb
Mv
R
R

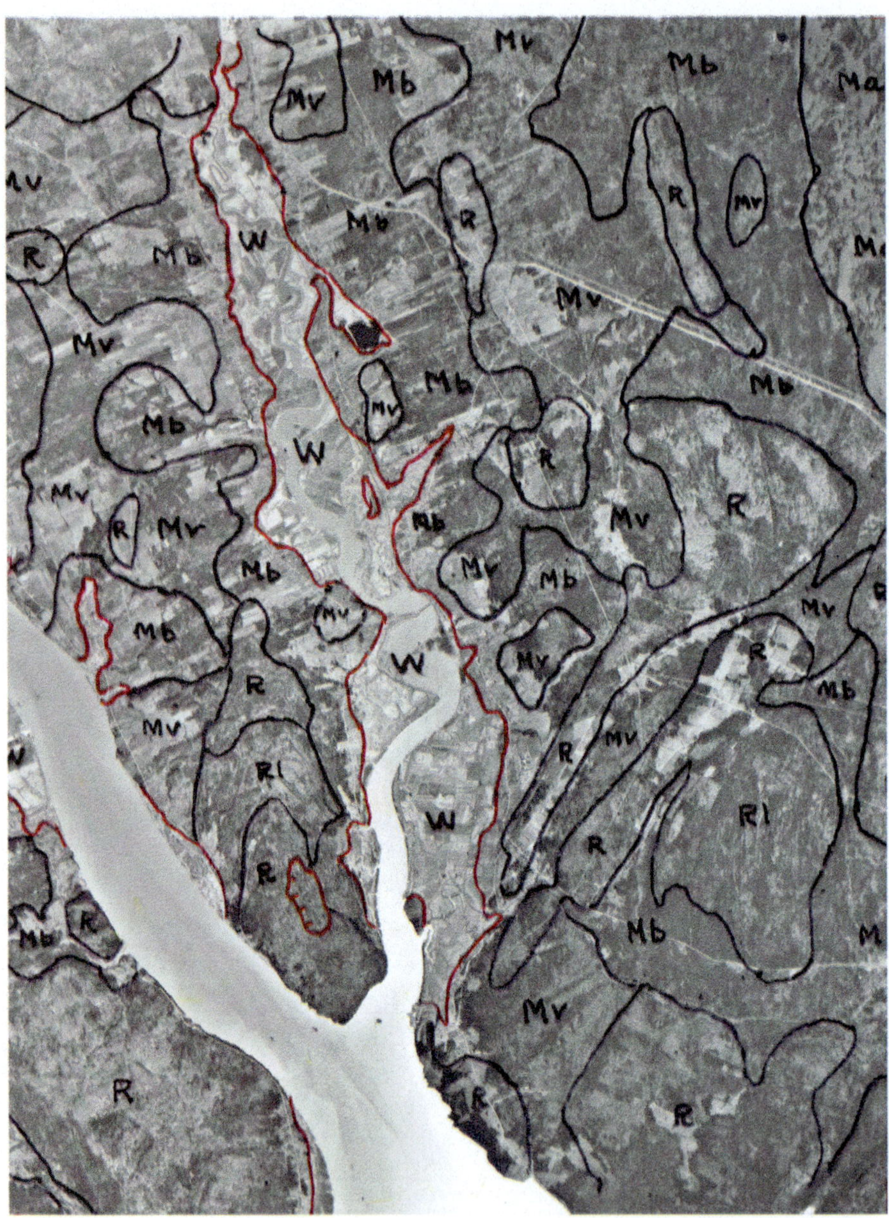

G7 Airphoto

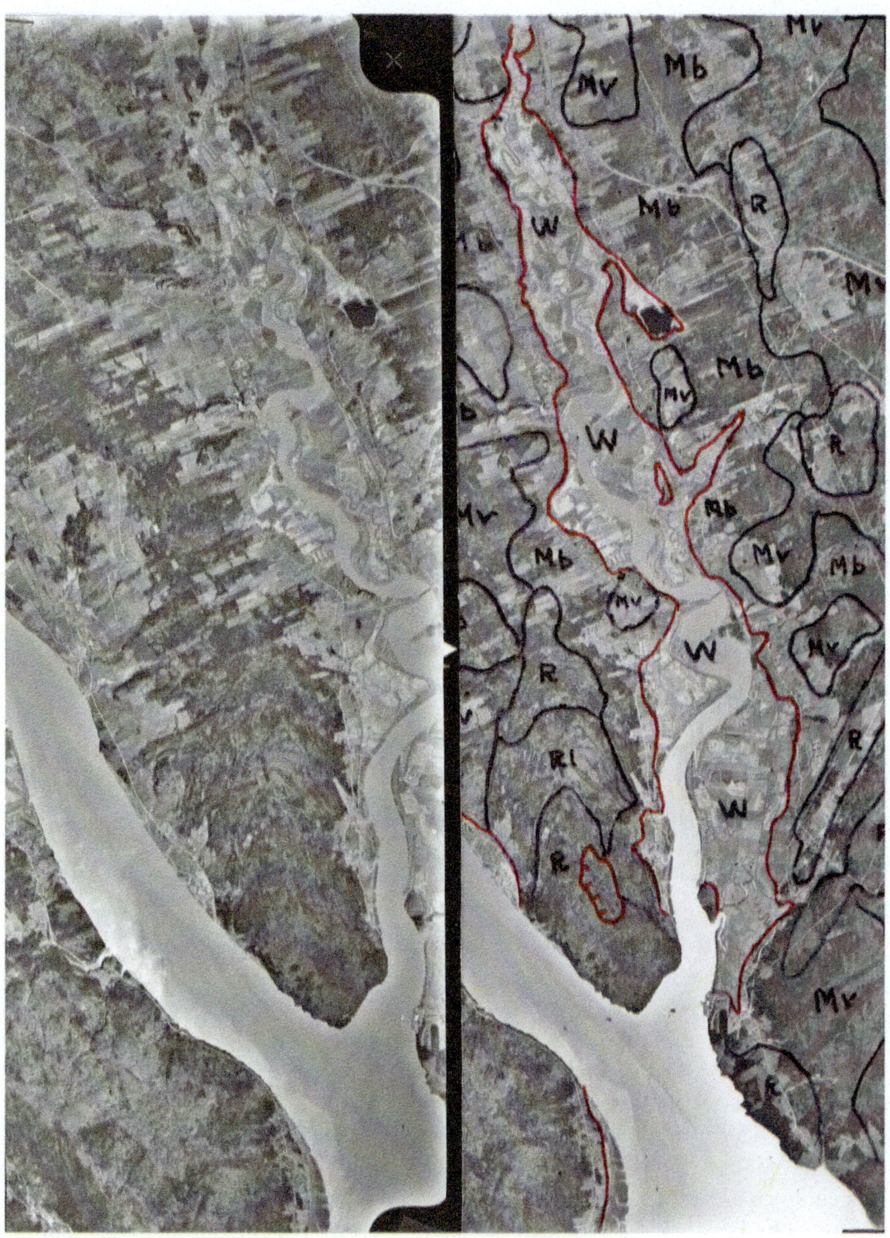

G7 Stereopair

Photo Interpretation

The 1:100,000 scale air photo of the site covers 350 km^2. Five terrain units are distributed in the area; three glacial deposits, marine deposits, and bedrock hills.

Glacial Deposits

These lie at 45–60 m above sea level.

Mb—ground moraine deposits occur on either side of Petitcodiac and Memramcook Rivers and adjacent to the off-site Tantramar Flats on the east.

Ma—This is an area of hummocky **ablation moraine** occurring in the northeast corner of the photo area.

Mv—veneer moraine is on higher ground near bedrock outcrops.

Marine Deposits

W—These are dyked and cultivated intertidal flats along the Memramcook River.

Bedrock

R—bedrock outcrops are mainly Lower Carboniferous Conglomerates.

R1—are **interbedded** shale and conglomerate beds.

Photo Source

Courtesy of National Air Photo Library, A21331, 43-44

Alpine Glacial Terrains

The eight Examples are largely the result of Late Quaternary deglaciation, 0.01 Ma.

Structures

- The French Alp Example is in a folded and thrusted flysch basin in the *Sub-Alpine Furrow* of the Alps between the Sedimentary High Alps on the east and the crystalline massif of *Pelvoux* on the west.
- The Yukon Examples are Lower Paleozoic carbonate rocks with peaks rising to 2,400 m elevation in the Interior System of the Cordillera between the Coast Mountains and the Rocky Mountains.
- The Labrador group of Examples are distributed from latitudes 57°–59° 45′ along 280 km of the entire length of the Torngat Mountains and Plateaux. The Examples are Paleoproterozoic thrusted and folded rocks which form the collision zone between the Archean blocks of the Rae Province in the west and the Nain Province to the east.

Figure G8: Fournel

Location 44°46′N, 06°26′E

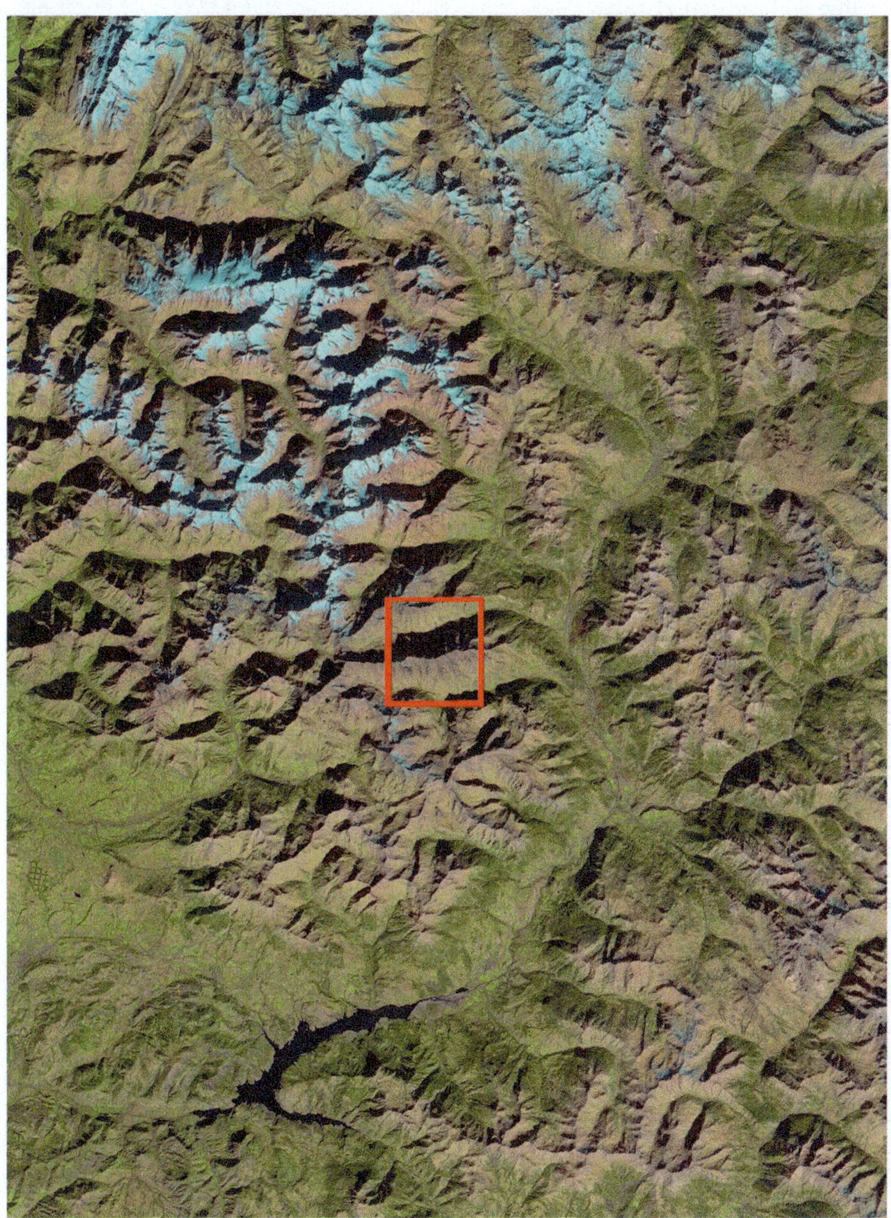

Landsat Image T/M 4-5, 16 October 2003

Regional Environment

The area is in the Pelvoux Hercynian massif of the French Alps. It adjoins Landslide Figure **E16** on the north. The massif has been glaciated and dissected into a confusion of ridge lines, peaks rising to 3,500 and 4,000 m with cirques, snow fields and glaciers. Half of the Example site is shadowed by a high ridge line. Snow-covered peaks and slopes are scattered over the north part of the scene.

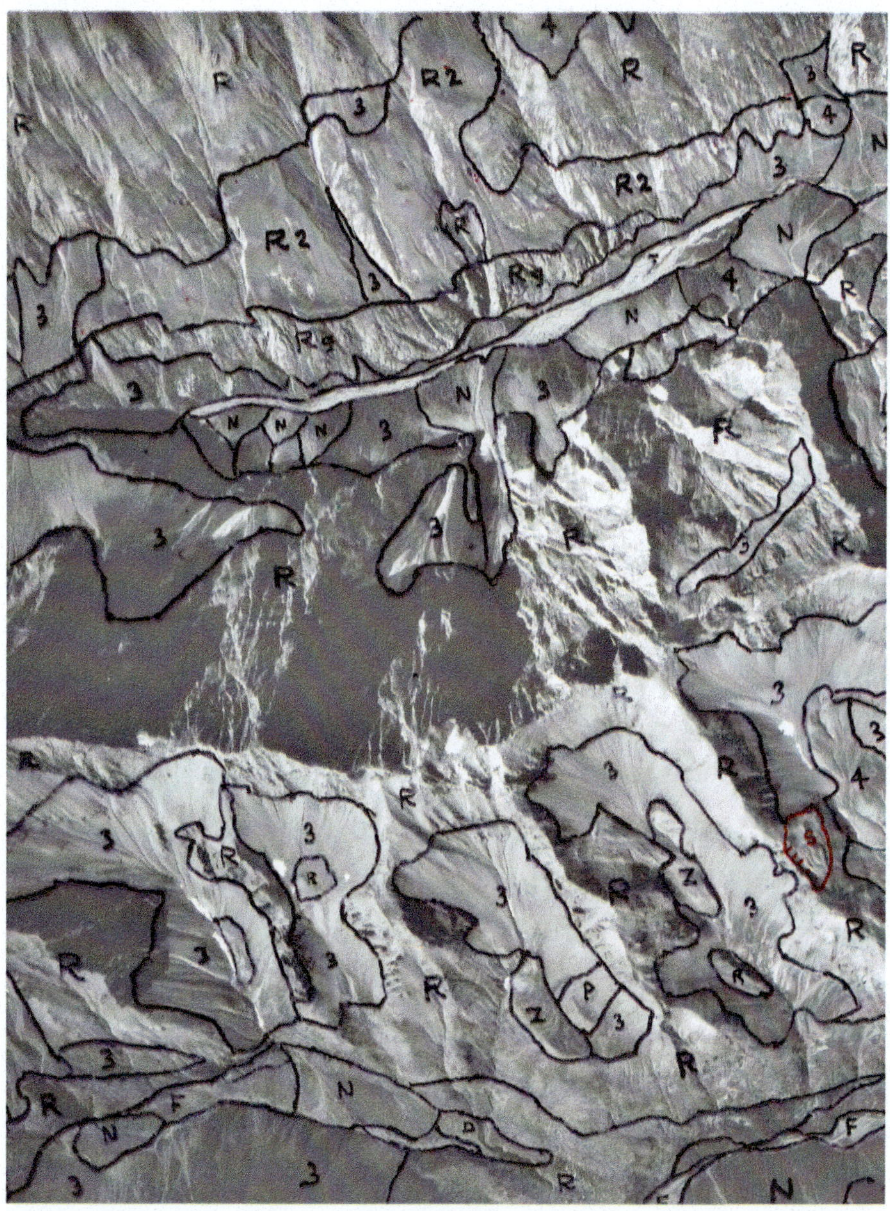

G8 Airphoto

G8 Stereopair

Photo Interpretation

The 1:30,000 scale air photo covers 35 km^2. Eight terrain types are distributed in five categories.

Bedrock Types

R—is interbedded Paleogene **sandstone and shale** overthrusted on **R2** sedimentary rocks.

R2—is Eocene **limestone**.

Rg—is basement **granite**.

Mass Movement

3—is **talus** lying in south-facing cirques and at the base of slopes near the stream valleys.

Glacial Deposit

4—are five local small occurrences of **ground moraine**.

Glaciofluvial Deposit

Z—are two small **kames** located on the margins of cirque talus deposits.

Alluvial Deposits

N—are **fans** at the base of slopes in the stream valleys.

F—is sand and gravel in the **braided stream** beds.

Photo Source

Figure G9: Bear River

Location 64°44′N, 134°05′W

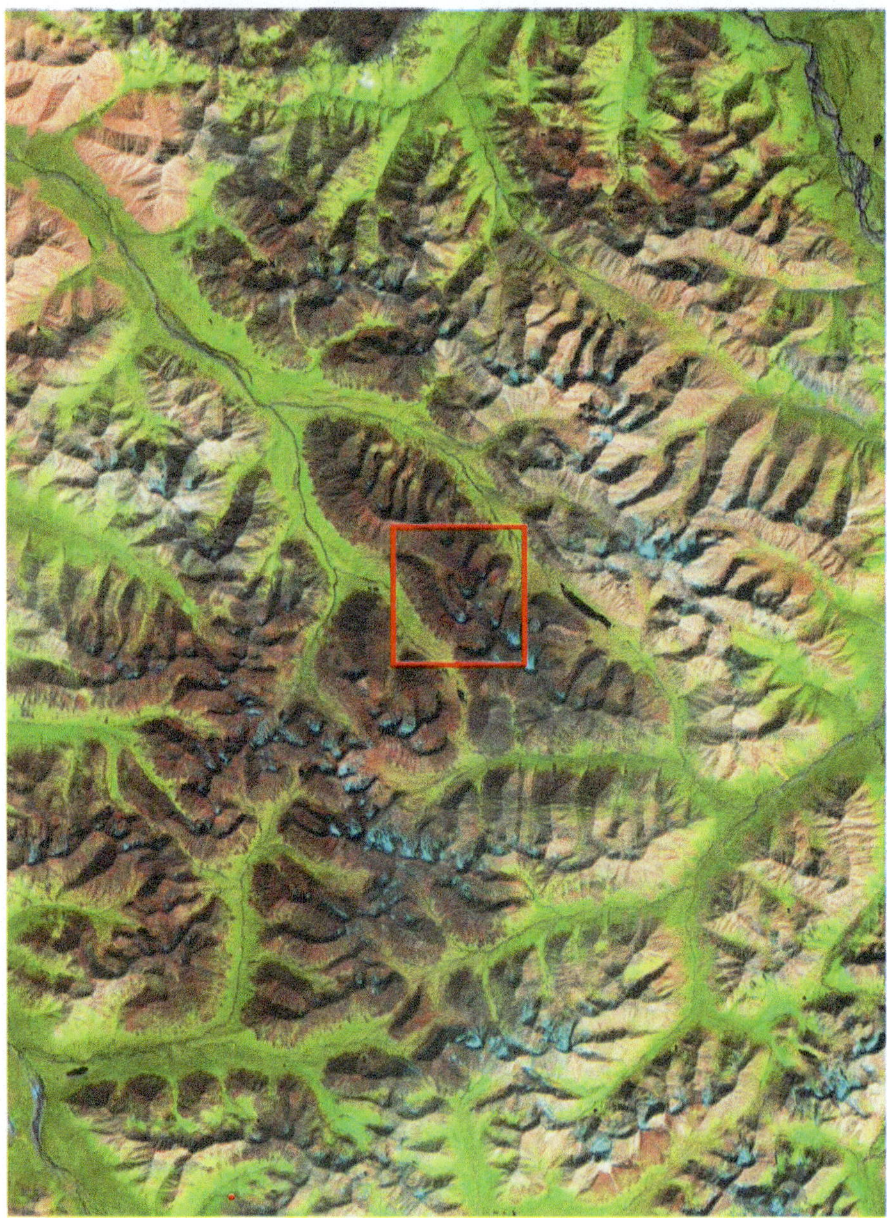

Landsat Image T/M 4-5, 14 July 1993

Regional Environment

The area is in numerous cirque topography with ridges above 2,000 m in the Selwyn Mountains of the Interior System of the Canadian Cordillera in east central Yukon.

Some of the north-facing rock glacier and debris-covered cirques are snow/ice covered.

R
3
F
G16.1
G16
R
R
R
Gt2
A
R
Gt4
G16.7
3
F
R
PTR
R
N
Zm2
3
3
R
3
R
Z m
R
3
R
R
3
(Zm4)
3
R
Zm2
3
R
3
Zm2
R
R
Z
3
Gt2
R
Gt2
R
3
R
R
3
R
F
Gt2
Gt2
R
3
R

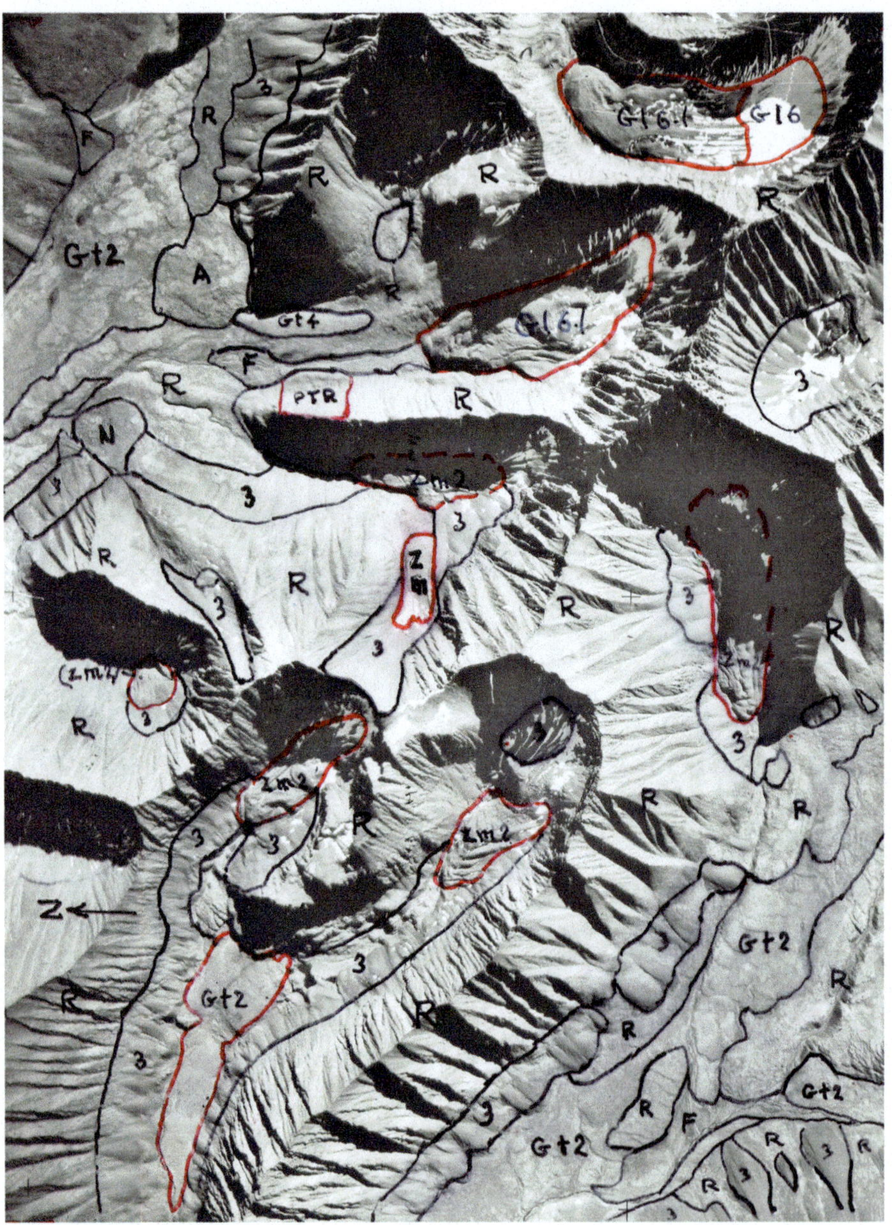

G9 Airphoto

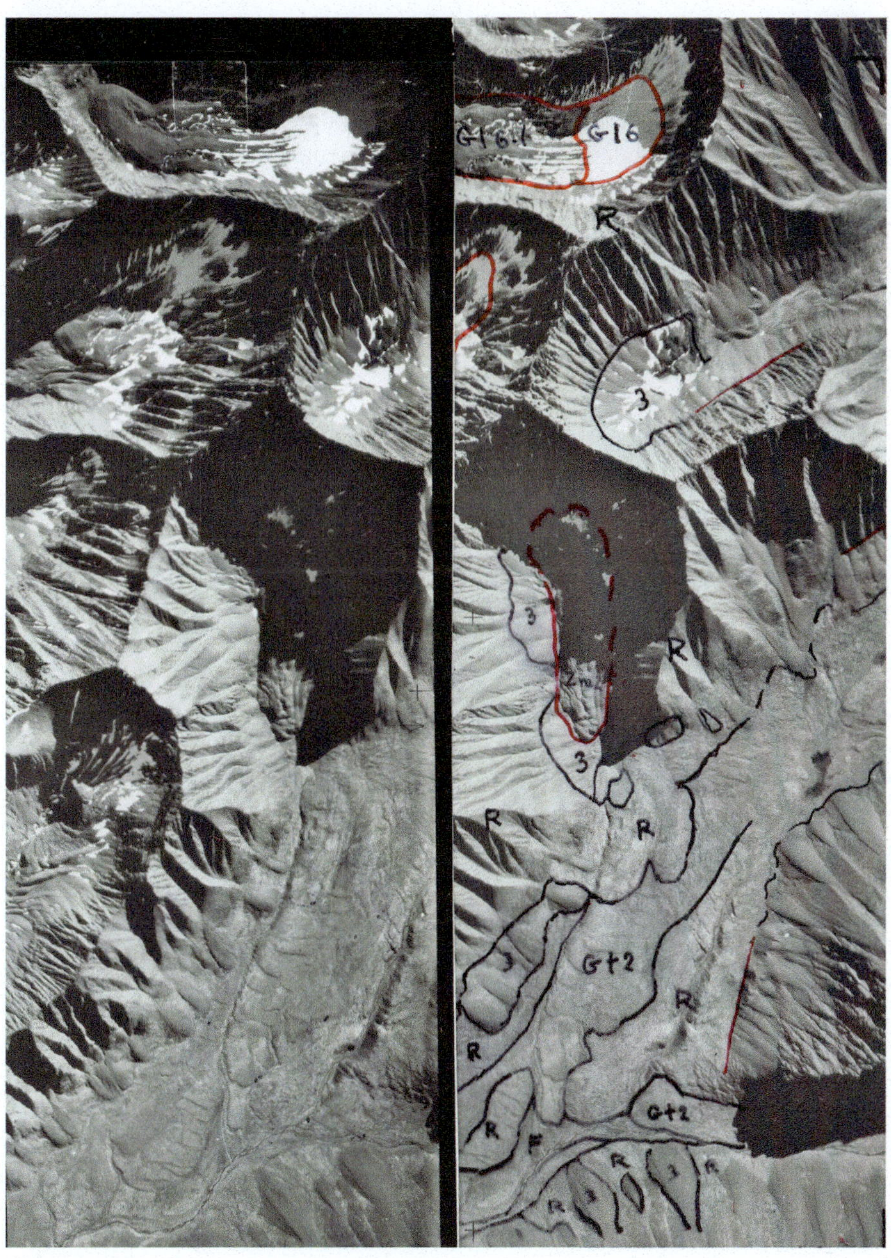

G9 Stereopair

Photo Interpretation

The 1:31,680 scale photo covers 35 km^2; *north arrow is in lower left.*

Multiple shadows obscure significant areas of north and west facing slopes in the high relief of the photo area. Ten units of five terrain categories occur in the site: three glacial; two periglacial; two of mass movements; two alluvial, and one bedrock type.

Glacial Deposits
Gl6—valley glacier
Gl6.1—2 glaciers covered with debris up to 1.5 m thick
Gt2—ground moraine

Periglacial Deposits
Zm1—gelifluction slopes
Zm2—rock glaciers

Mass Movements
A—debris avalanche
3—talus

Alluvial Deposits
N—fan
F—sand gravel stream beds

Bedrock
R—folded Paleozoic interbedded sediments

Photo Source

Courtesy of National Air Photo Library, A12283, 142–143

Figure G10: Carmacks

Location 61°58′N, 136°25′W

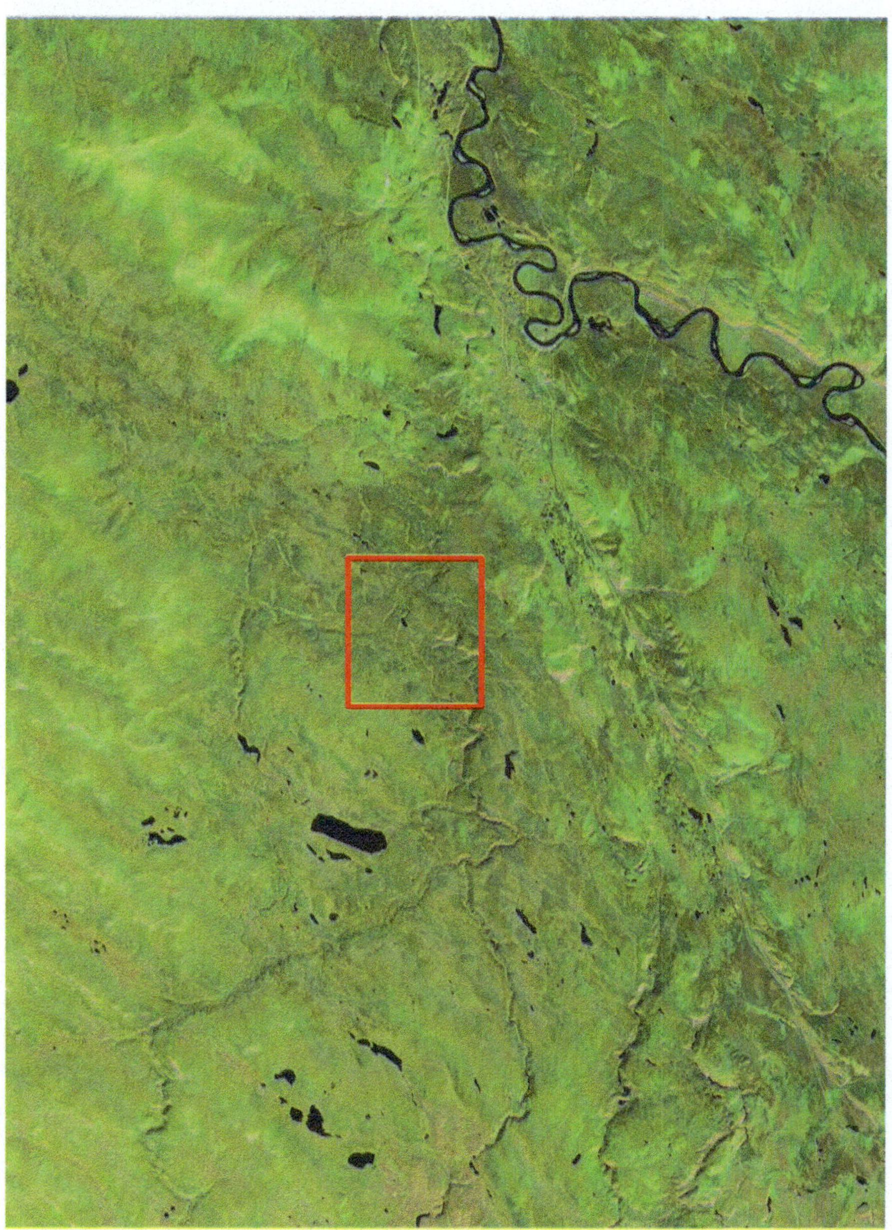

Landsat Image T/M 4-5, 11 August 2003

Regional Environment

The area is in the Lewes Plateau of the Interior System of southwest Yukon.

Structure

The Example is located in a broad depression of mainly sedimentary rocks between the west ranges of the crystalline Pelly Mountains on the east and the dissected and faulted metamorphic rocks of the Klondike Plateau on the west.

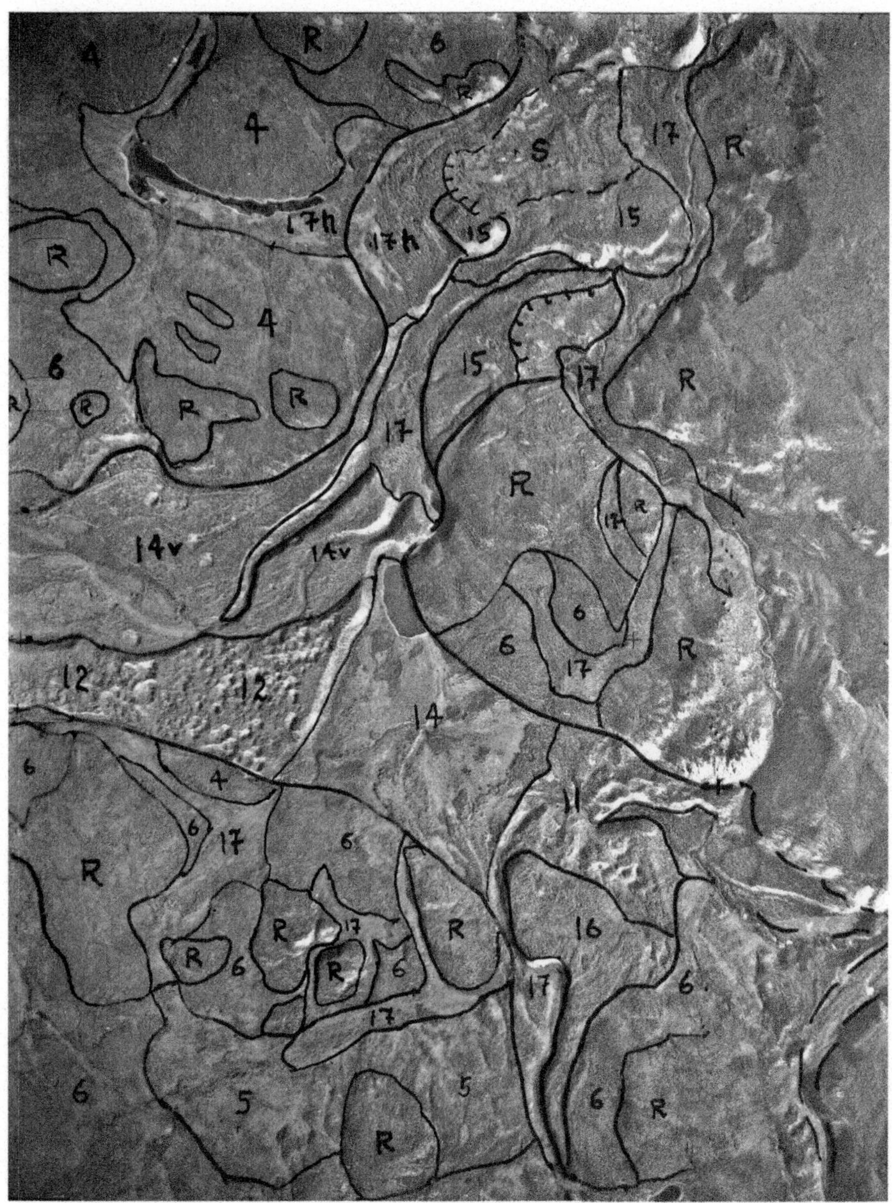

G10 Airphoto

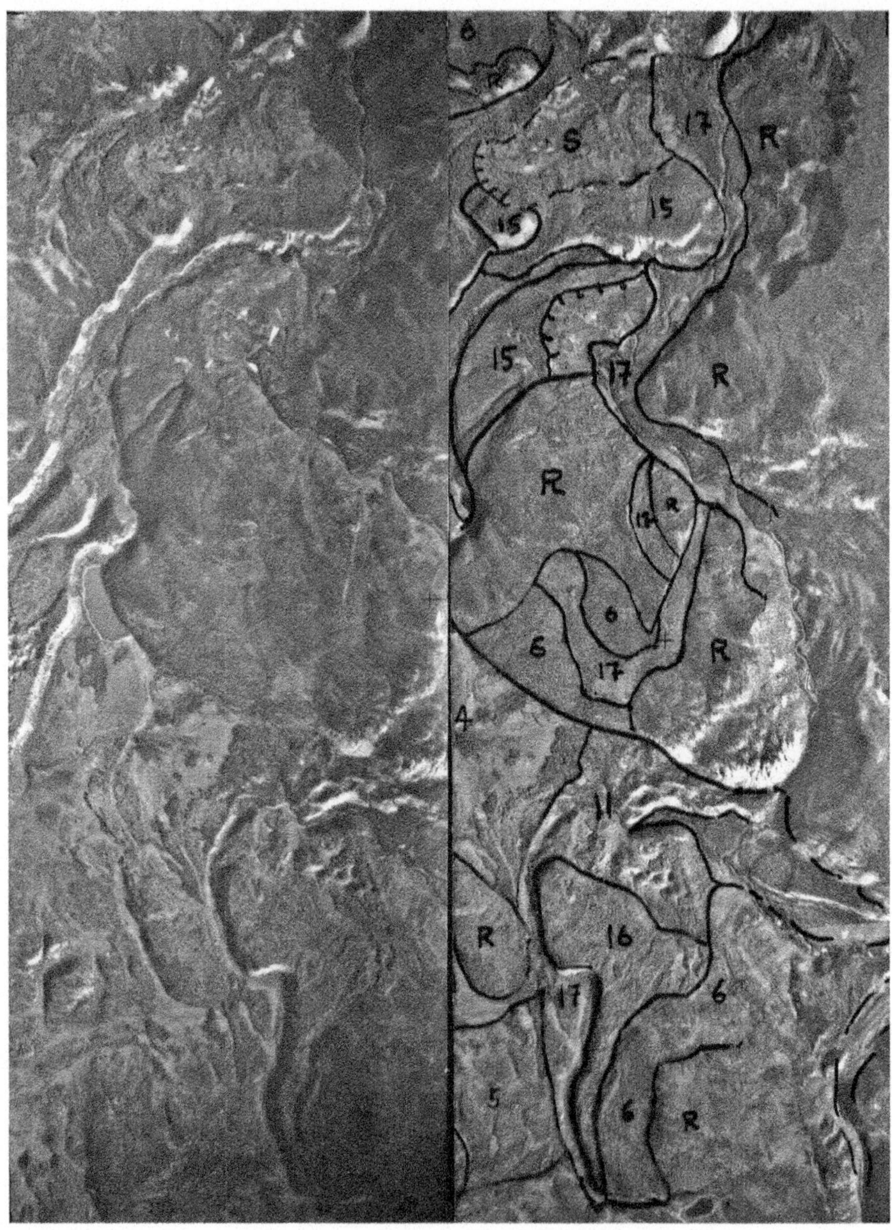

G10 Stereopair

Photo Interpretation

The 1:40,000 scale photo covers 56 km^2 at a general elevation of 850 m.

Thirteen units of four terrain categories occur in the site; three glacial, eight fluvioglacial, one mass movement, and one bedrock type.

Glacial Deposits
4—ground moraine
5—reworked moraine
6—veneer moraine

Glaciofluvial Deposits
11—kame
12—pitted kame terrace
14—outwash plain at 810 m elevation
14v—outwash valley train at 840 m elevation
15—terrace
16—delta
17—valley spillway
17h—elevated spillway

Mass Movement
S—debris slide in terrace sediments

Bedrock Units
R—Upper Cretaceous volcanic rocks at elevations ranging from 740 to 860 m

Photo Source

Courtesy of National Air Photo Library, A19537, 12-13

Figure G11: Mt Blow-Me-Down

Location 58°47′N, 63°00′W

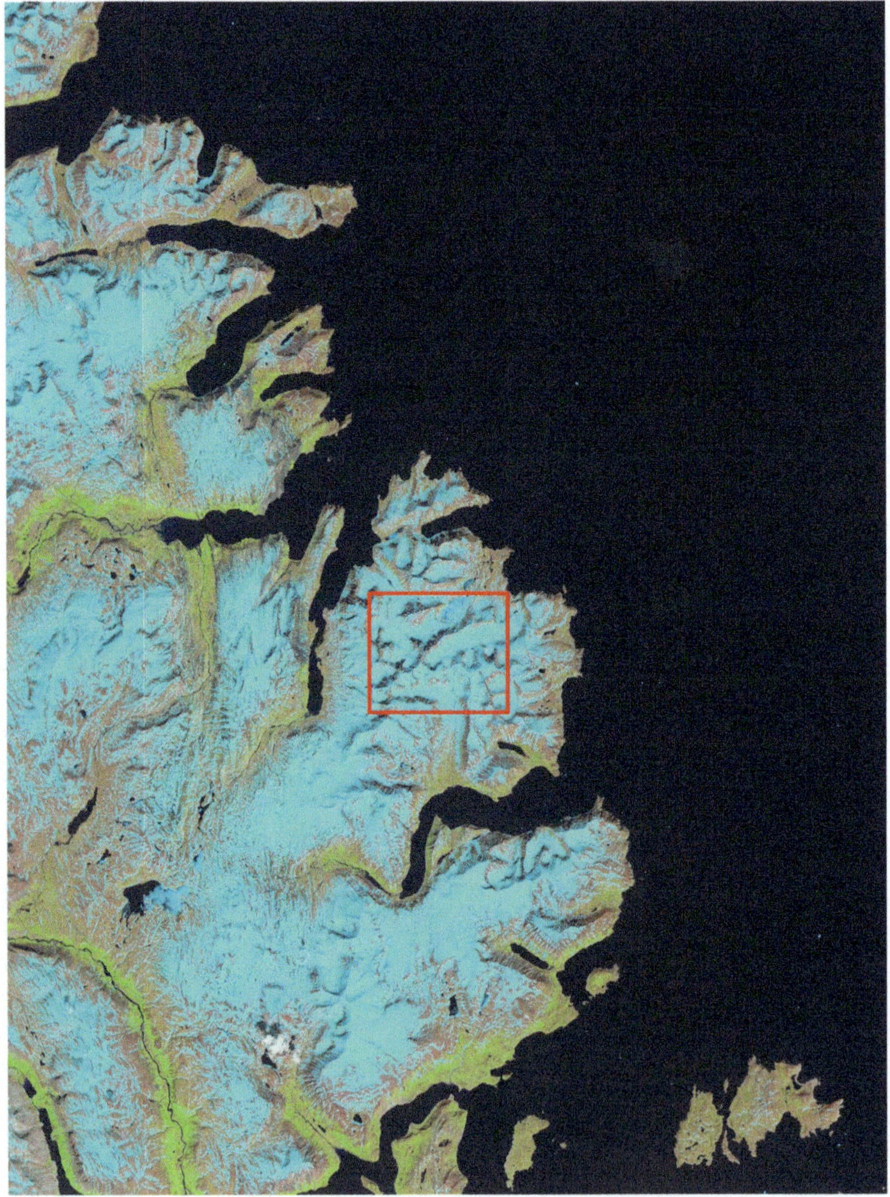

Landsat Image T/M 4-5, 06 July 2010

Regional Environment

The area is a snow-covered fjord-bounded peninsula of alpine glacial erosion in the northern Torngat Mountains, Labrador.

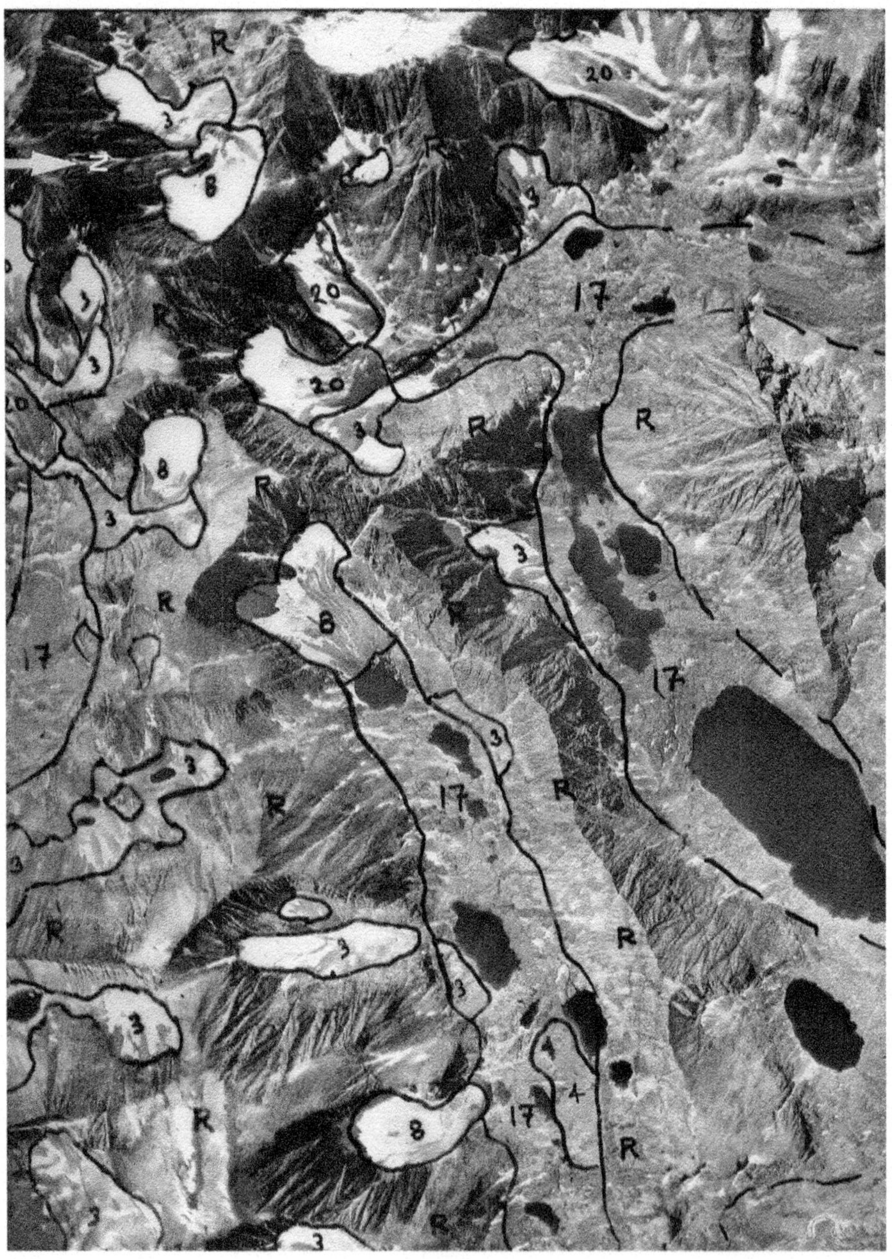

G11 Airphoto

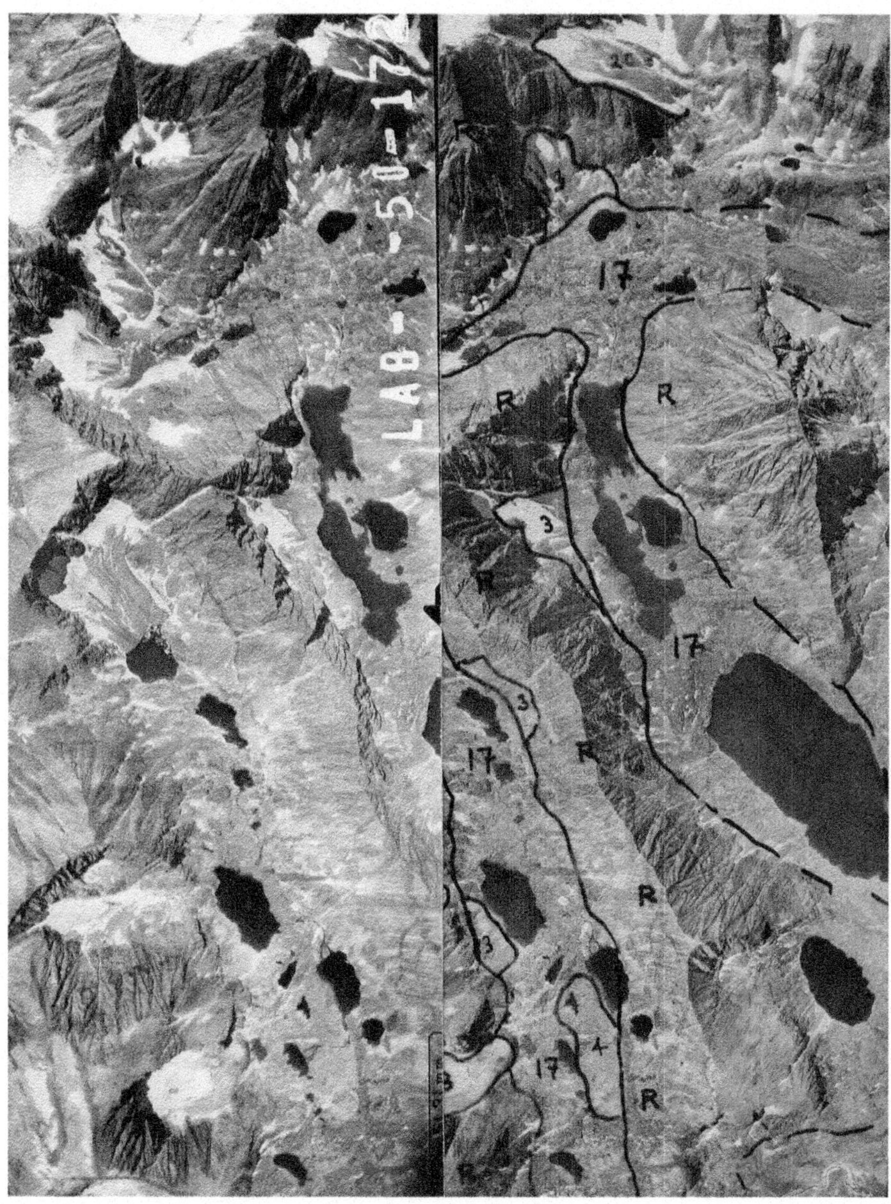

G11 Stereopair

Photo Interpretation

This 1:40,000 scale photo covers 56 km^2 of terrain, *part of a north arrow is in the upper left*. Six units of five terrain categories occur in the site area.

Mass Movements

3—generally snow-covered **talus** sheets

Glacial Deposits

4—a single 1 km long by 300 m wide deposit of **reworked moraine** lies in a 700 m wide spillway.

8—four deposits of 700–900 m long, generally snow-covered, **marginal moraine** in cirque basins.

Glaciofluvial Deposits

17—three 1 km wide **spillway channels** in glaciated trough-sloped valleys drain to the sea in north, east and south directions respectively.

Periglacial Deposits

20—5,600 m to 1 km long snow-covered **rock glaciers** head in cirques in the southwest quarter of the photo.

Bedrock Units

R—slopes and ridges of Archean gneiss rise to 700–890 m elevations.

Photo Source

Courtesy of National Air Photo Library, LAB-50, 171-172

Reference

See Figure G15

Figure G12: Eclipse River

Location 59°43 N, 64°44′W

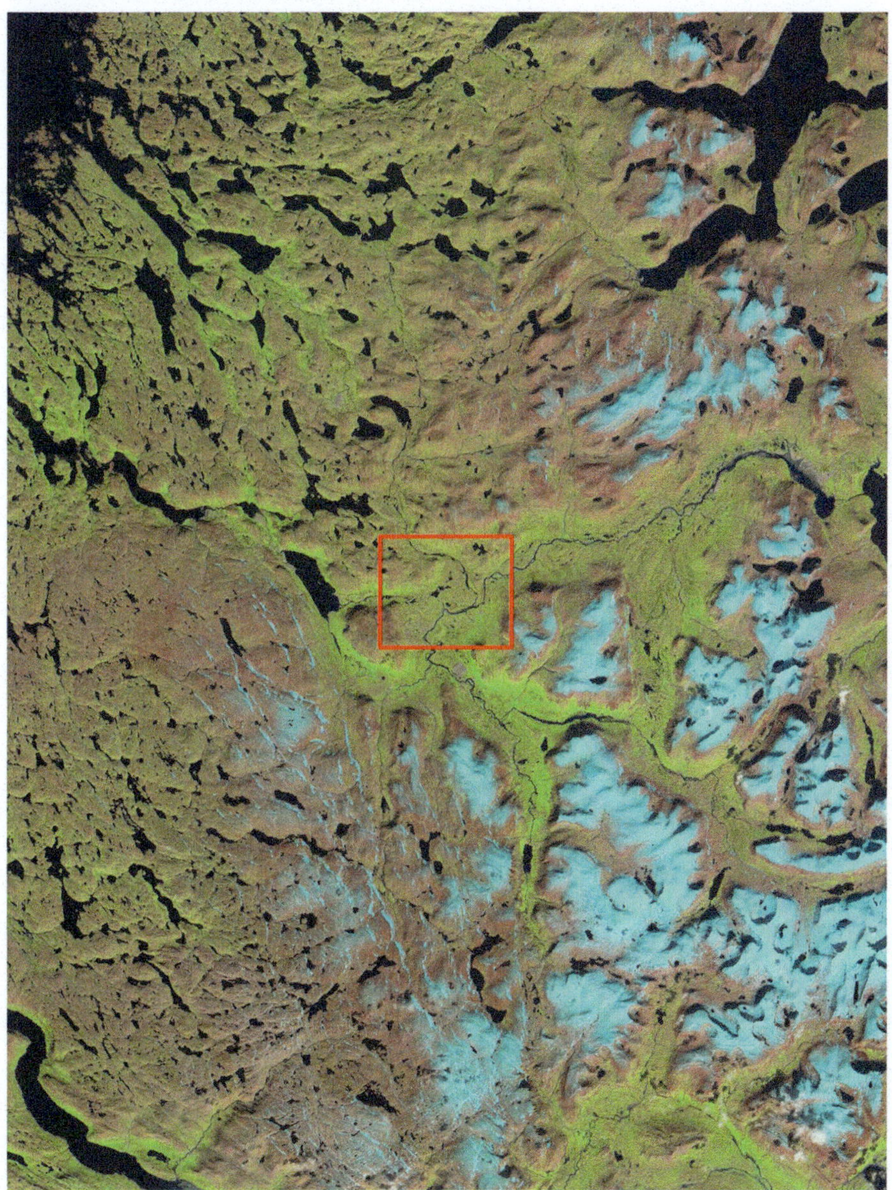

Landsat Image T/M 4-5, 05 September 2004

Regional Environment

The area is 140 km northwest of Figure G11 in a locality 35 km inland from the coast in the northern extremity of the Torngat Mountains, Labrador.

Structure

The area is in Proterozoic gneiss. North striking faults paralleling the coast are not detectable on Landsat imagery.

Physiography

The site is a 90 m elevation basin-shaped enlargement of the valley of the Eclipse River. At 10 ka time of deglaciation 8 km broad and 600–270 m elevation glaciers from the Ungava watershed crossed the Torngats along the Eclipse Valley and extended as outlet glaciers to the Labrador Sea at Eclipse Channel (Clark 1984). Today the valley is occupied by 1–2 km broad outwash valley train deposits over a length of 30 km from the Example site, leaving the narrow Eclipse River as a misfit stream.

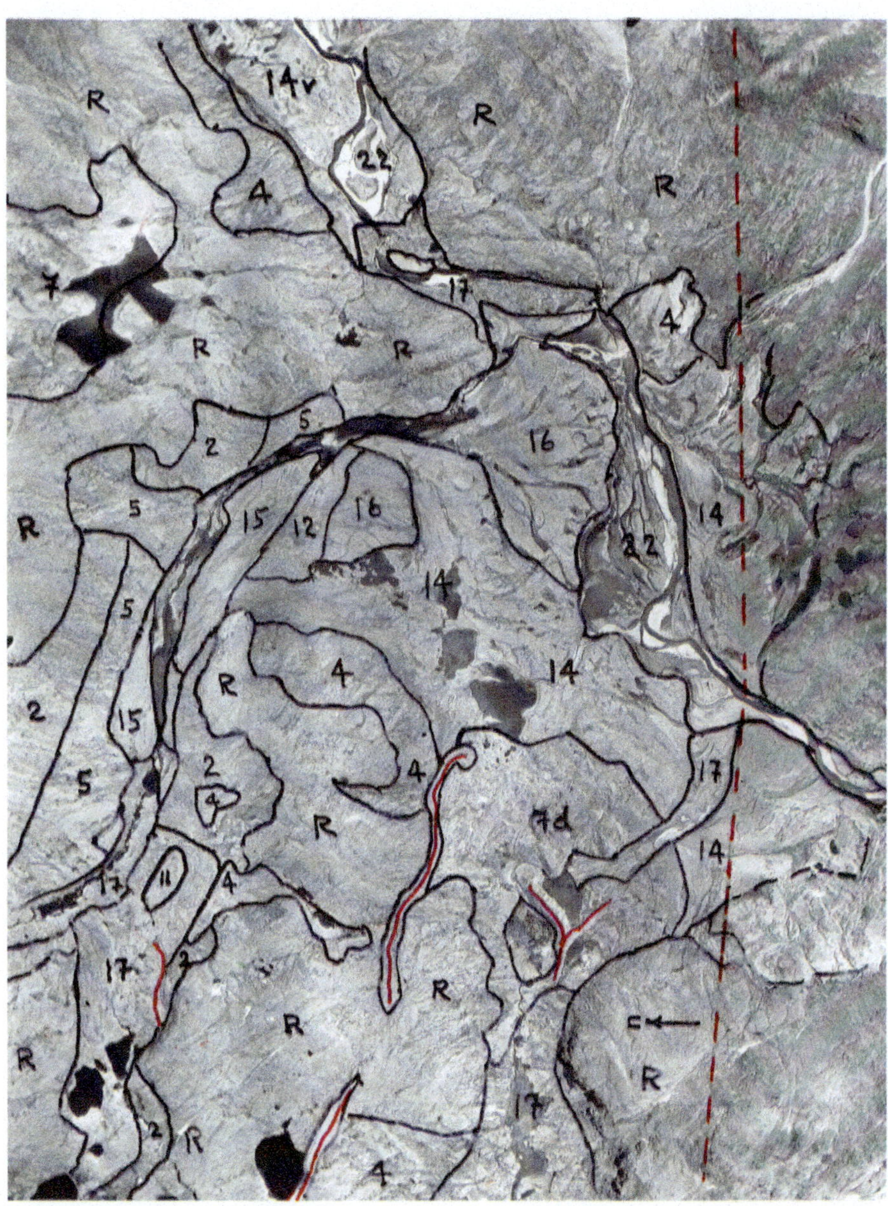

G12 Airphoto

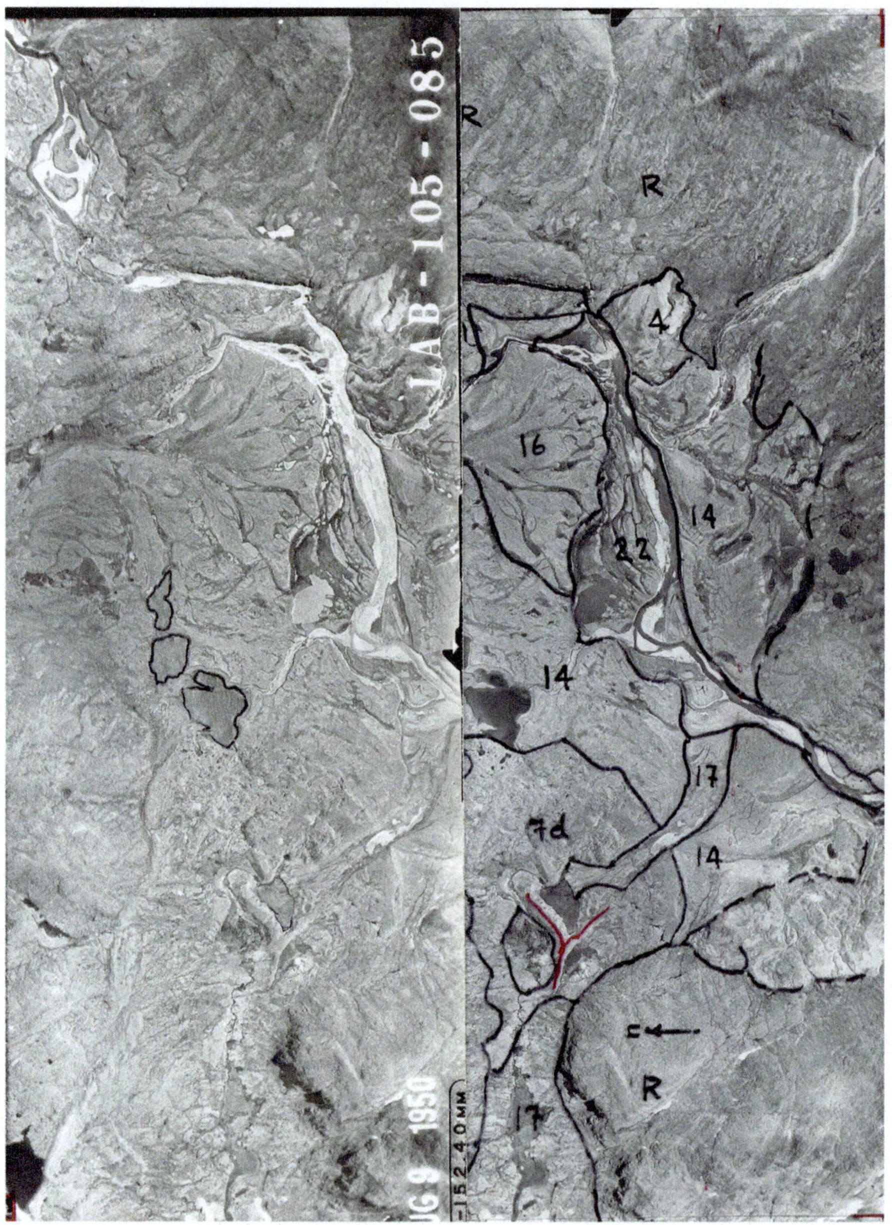

G12 Stereopair

Photo Interpretation

The 1:40,000 scale photo covers 56 km^2. Thirteen terrain units in four categories occur in the site.

North arrow is in lower right.

Glacial Deposits

4—five units of **ground moraine** are located adjacent to bedrock outcrops
5—reworked moraine
7—a single occurrence of **hummocky moraine** lies next to the lake in upper left
7d—a 2 km wide deposit of **stagnant ice moraine** lies in the basin

Glaciofluvial Deposits

10—four **esker** ridges are traced in red
11—a single **kame** mound lies in a spillway channel in lower left
12—a larger ice-contact **kame terrace** lies slightly above a proglacial terrace
14—an extensive 3 km long by 1.5 km wide **outwash plain** lies in the center of the site basin
15—two **proglacial terraces** lie along a local stream channel
16—a 1,700 m wide proglacial delta lies at the mouth of the terrace-bordered stream

Alluvial Deposits

22—the braided stream channel is part of the Eclipse River system

Bedrock Units

R—gneiss hills enclose the site basin at elevations ranging from 120 to 390 m
2—rock free slope

Photo Source

Courtesy of National Air Photo Library, LAB-105, 084-085

Reference

See Figure G15

Figure G 13: Laura Lake

Location 57°12′N, 62°16′W

Landsat Image T/M 4-5, 31 August 2004

Regional Environment

The scene covers 3,250 km^2 of central Labrador. A pink zone of ice flow markings in veneer ground moraine are distinct. Two northwest striking faults on the south side of Okak Bay are expressed by vegetated parallel valleys. An arrow points to a bare arcuate ridge detectable at the head of 20 m elevation Tasiuyak Lake. Just north of the site, is a group of 80–90 m elevation glaciomarine raised beaches

Physiography

The example area is in the inner basin of a group of glacially scoured low lake basins about 15 km inland from the coast infilled with post-glacial deposits. The deposits in the basin range from 120 to 260 m elevation.

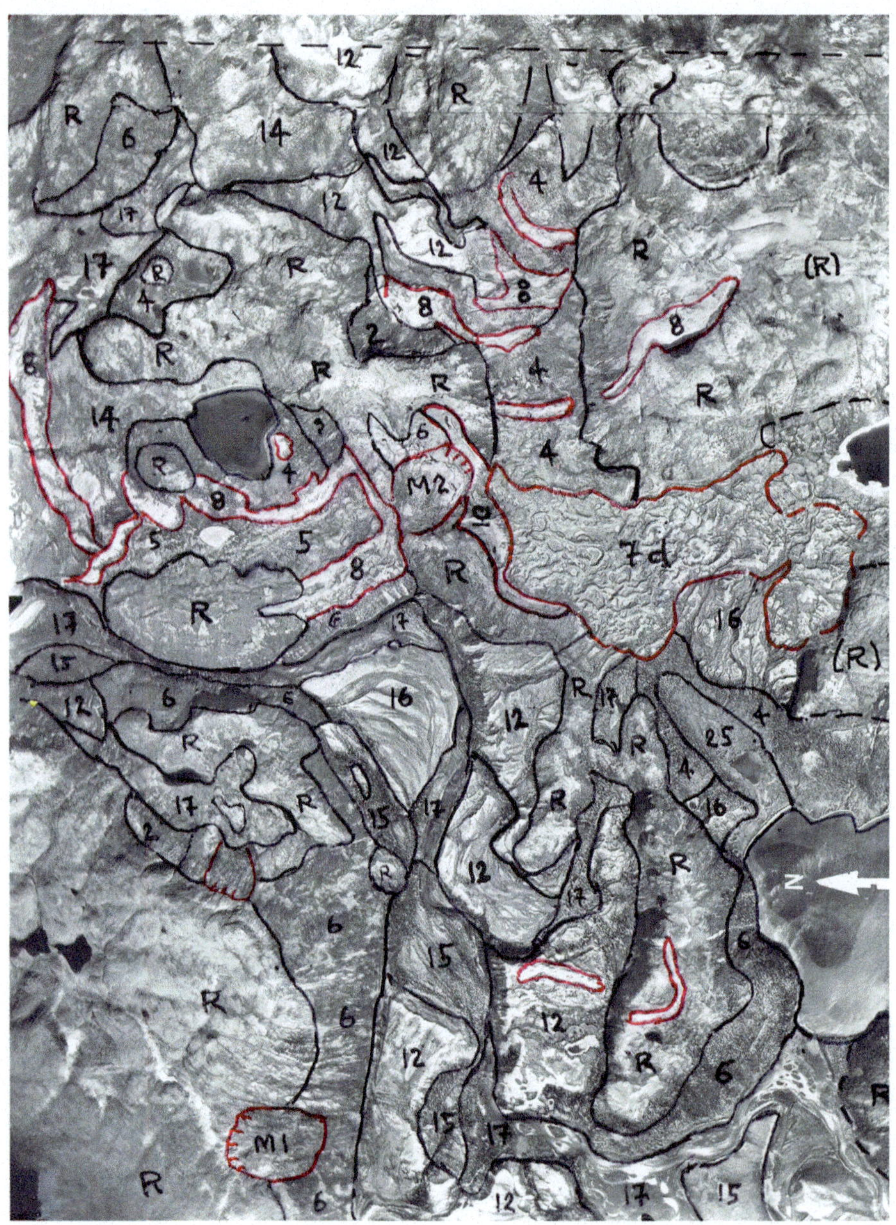

G13 Airphoto

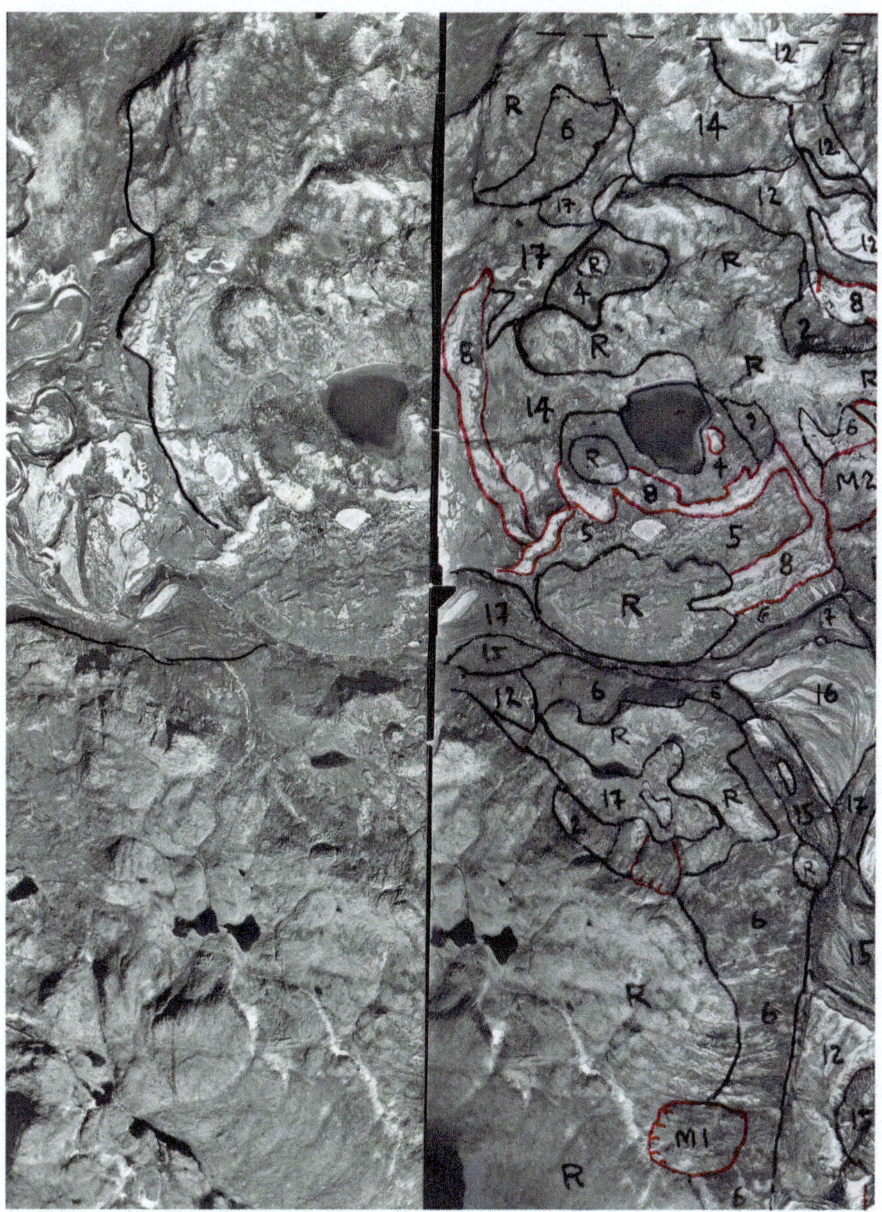

G13 Stereopair

Photo Interpretation

The 1:40,000 scale photo covers 56 km^2. Fifteen terrain units in five categories occur in the site.

Glacial Deposits

4—there are five occurrences of **ground moraine** of various sizes in low sites throughout the area

5—a single occurrence of **reworked moraine** is associated with the marginal moraine (8) complex near the small lake

6—veneer moraine deposits are located on the slopes of bedrock outcrops

7d—a single deposit of **stagnant ice moraine** with characteristic hummocky collapse relief lies in a 260 m elevation site on the south side of the basin

8—a complex pattern of arcuate bare-surfaced **marginal moraines** encircle the small lake; a group of others occur southward. There is a possible association with raised beaches.

Glaciofluvial Deposits

10—a 2 km long esker lies between the 7d stagnant ice moraine and the rock ridge

12—a group of large near-bare *ice-contact* **kame terraces** lie above *proglacial* terraces in the depression of the east and north flowing stream

14—an **outwash plain** lies between the lake and a large marginal moraine

15—this is a flow-marked *proglacial* **terrace** lying below the kame terraces in the east-flowing stream valley

16—a typical **delta** deposit has developed at the constricting bend of the east and north-flowing stream

17—the bed of the through-flowing stream lies in a **spillway channel**

Galciolacustrine Deposit

25—a narrow deposit of lacustrine sediments lies beyond the north shore of the large lake

Mass Movements

M1—is a small **debris slide** in a pocket of colluviums on the slope of a bedrock hill

M2—is a rock slide in a rock hill mass

Bedrock Units

R—local bedrock hills are Proterozoic anorthosite

Photo Source

Courtesy of National Air Photo Library, LAB-52, 115-116

Figure G14: Siugak Brook

Location 57°42′N, 62°44′W

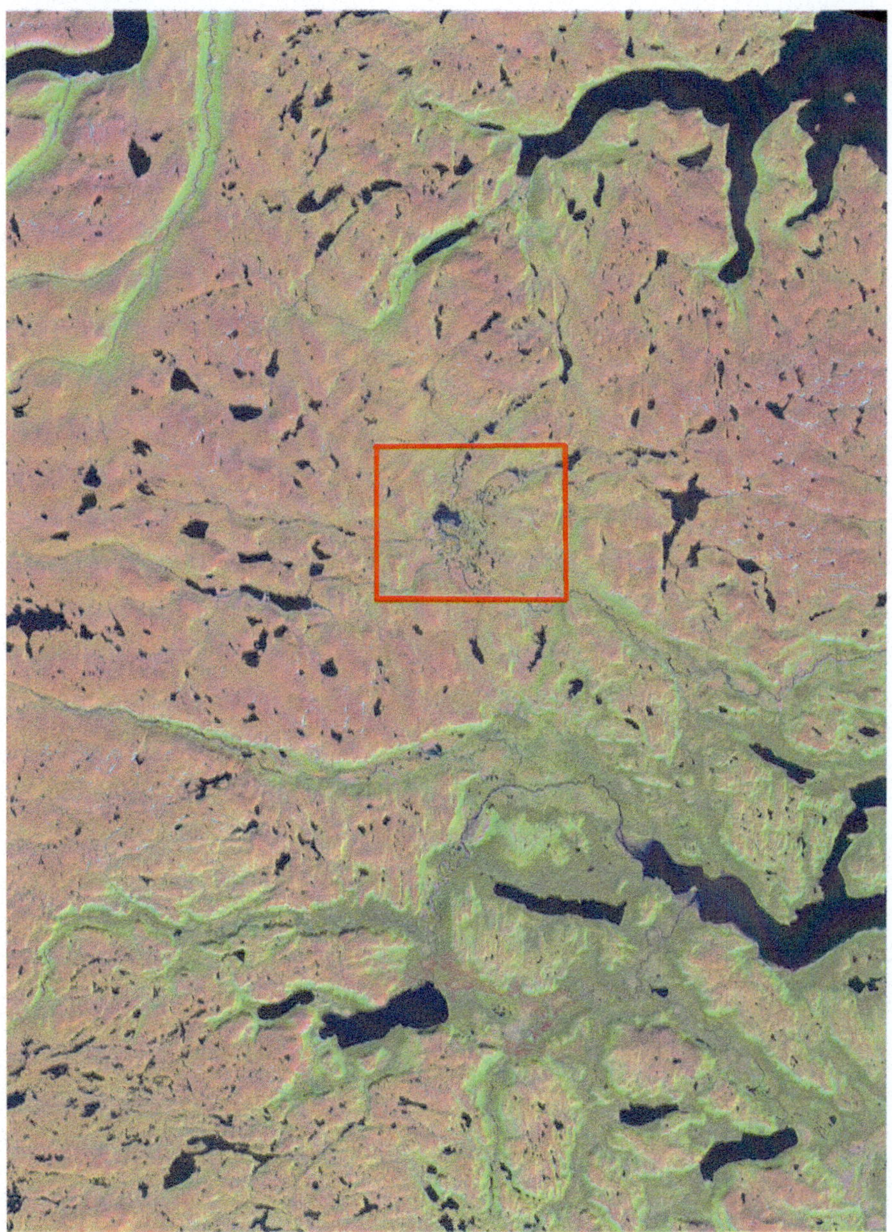

Landsat Image T/M 4-5, 27 July 2006

Regional Environment

The area is 35 km north of Figures G13 and G15 in central Labrador.

Physiography

The Example site is a ponded basin at 200 m elevation amid surrounding hills of plutonic rocks that rise to 400–500 m. The basin is fed by four tributary streams from north and west and drains southeastward by a fifth stream 35 km to Okak Bay.

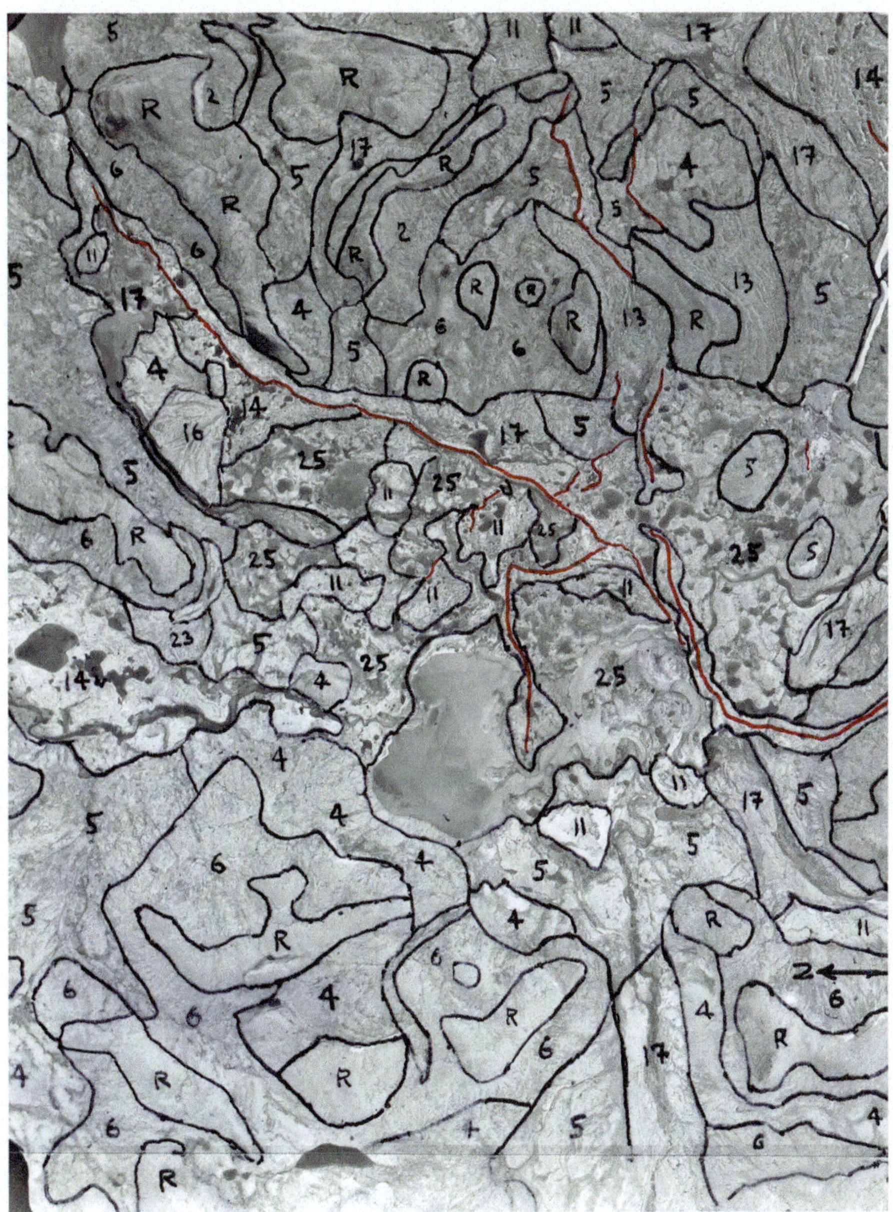

G14 Airphoto

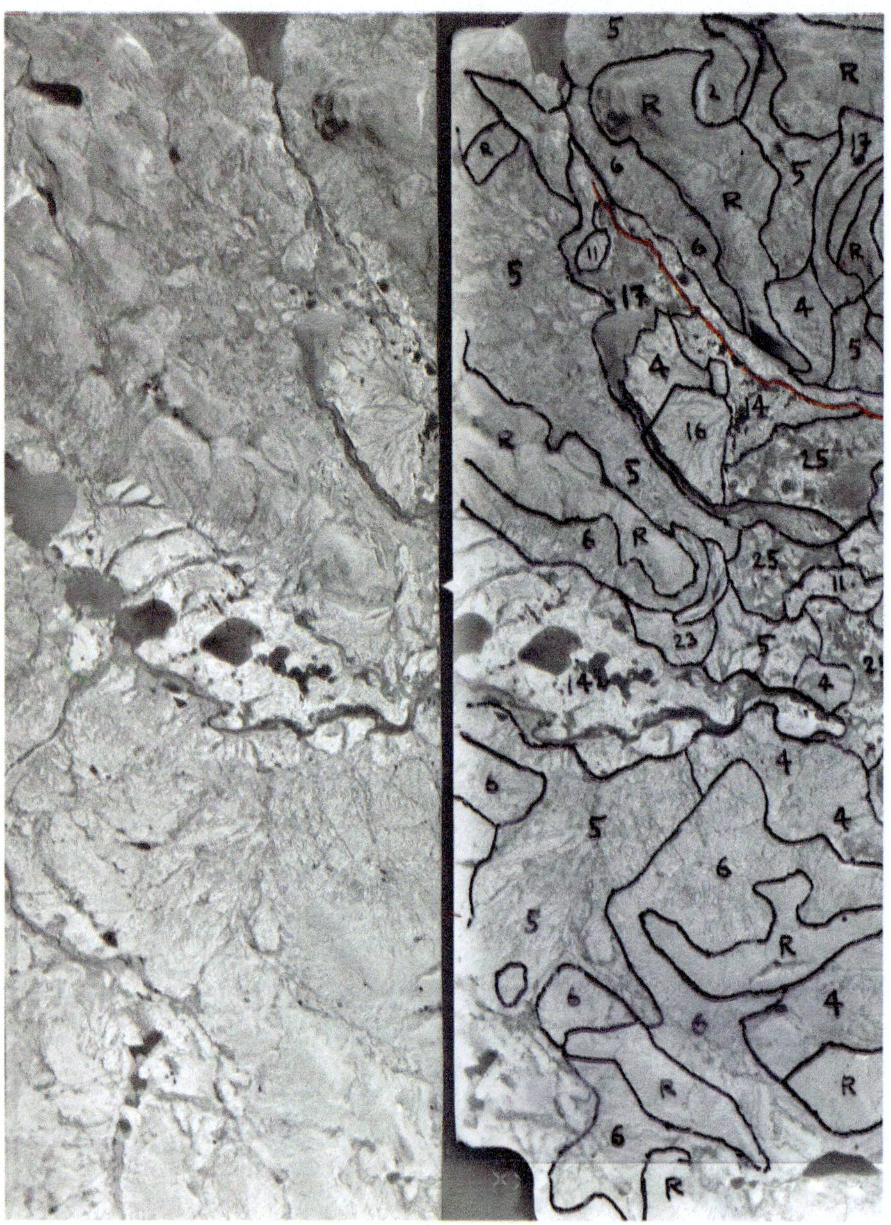

G14 Stereopair

Photo Interpretation

The 1:60,000 scale air photo covers 117 km^2, 11 units in four categories occur in the site. *A north arrow is in lower right.*

Glacial Deposits
4—ground moraine
5—reworked moraine
6—veneer moraine

Glaciofluvial Deposits
10—an extensive system of eskers traced in red are located in the basin
11—five small low kames lie adjacent to lacustrine deposits; two larger units are in a spillway leading to the outwash plain and another is located along a spillway channel in the southwest
13—two marginal meltwater channels are along the margin of small bedrock hills
14—a pitted outwash plain and a short esker lie in the bend of a stream in a spillway channel in the southeast corner of the photo. A larger, bare appearing, also pitted, occurrence is at the north edge of the photo.
16—a single delta is in a spillway channel parallel to the esker
17—three spillway channels lead into the basin. A fourth leads to the outwash plain in the southeast.

Glaciolacustrine Sediments
25—these sediments surrounding the lake with layers of peat are the dominant unit of the site

Bedrock Units
R—all outcrops are Proterozoic granite

Photo Source

Courtesy of National Air Photo Library, A18299, 76-77

Reference

See Figure G15

Figure G15: Umiakovik Lake

Location 57°22′N, 62°41′W

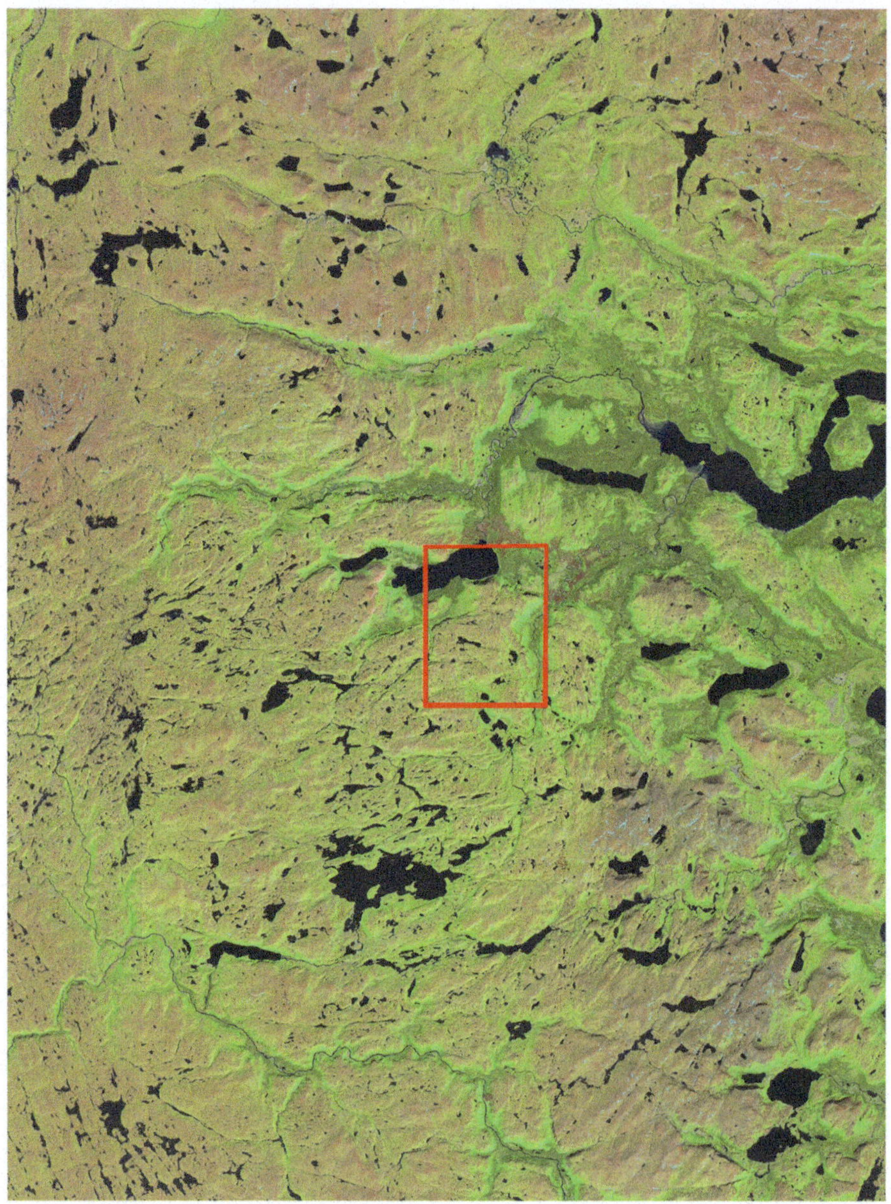

Landsat Image T/M4-5, 27 July 2006

Regional Environment

The area is 25 km west of Okak Bay north Labrador coast.

Structure

The example is in a Proterozoic granite batholiths near the central Labrador coast.

Physiography

The area is in a region characterized by rugged block faulted bedrock terrain at elevations ranging from 500 to 670 m with east west oriented strong fractures or fault lineaments. Local drainage systems all lead to Okak Bay. Lake elevation is 29 m.

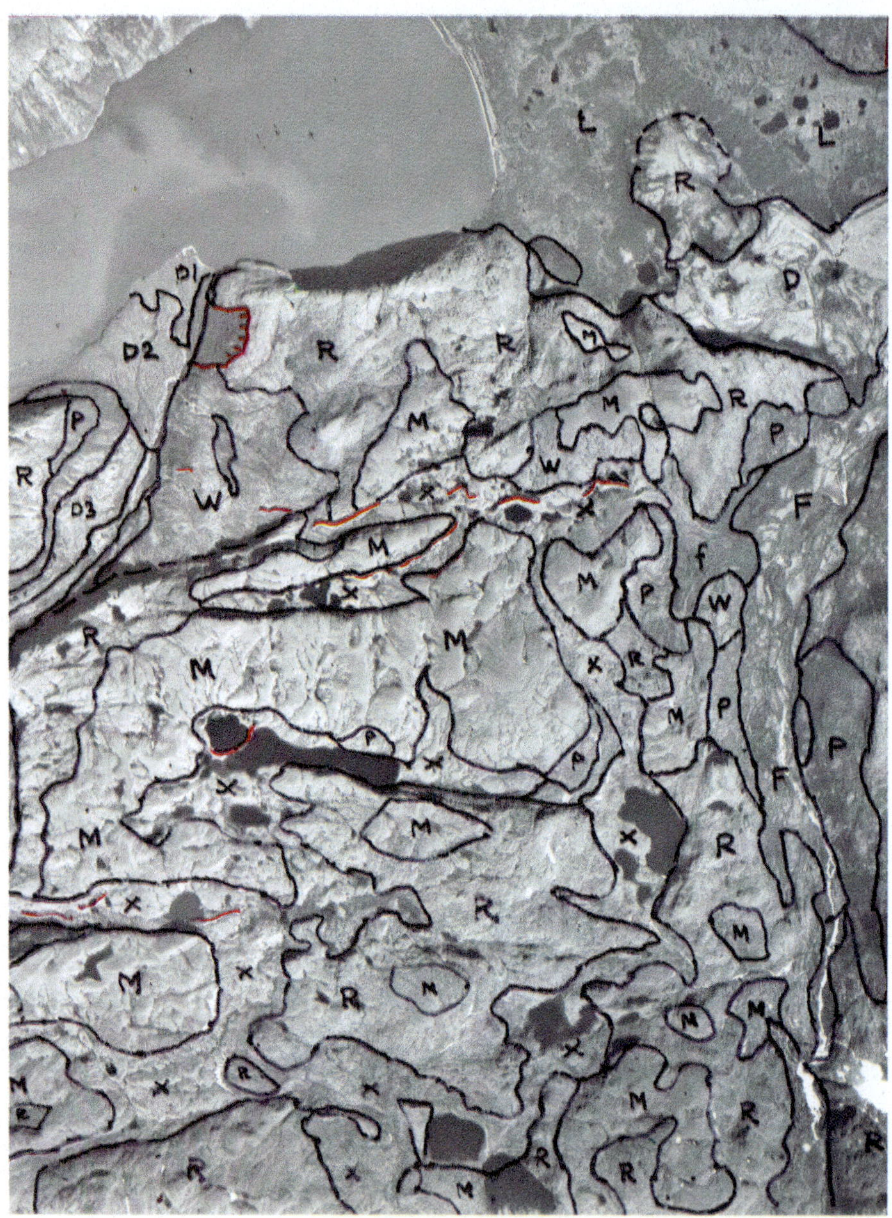

G15 Airphoto

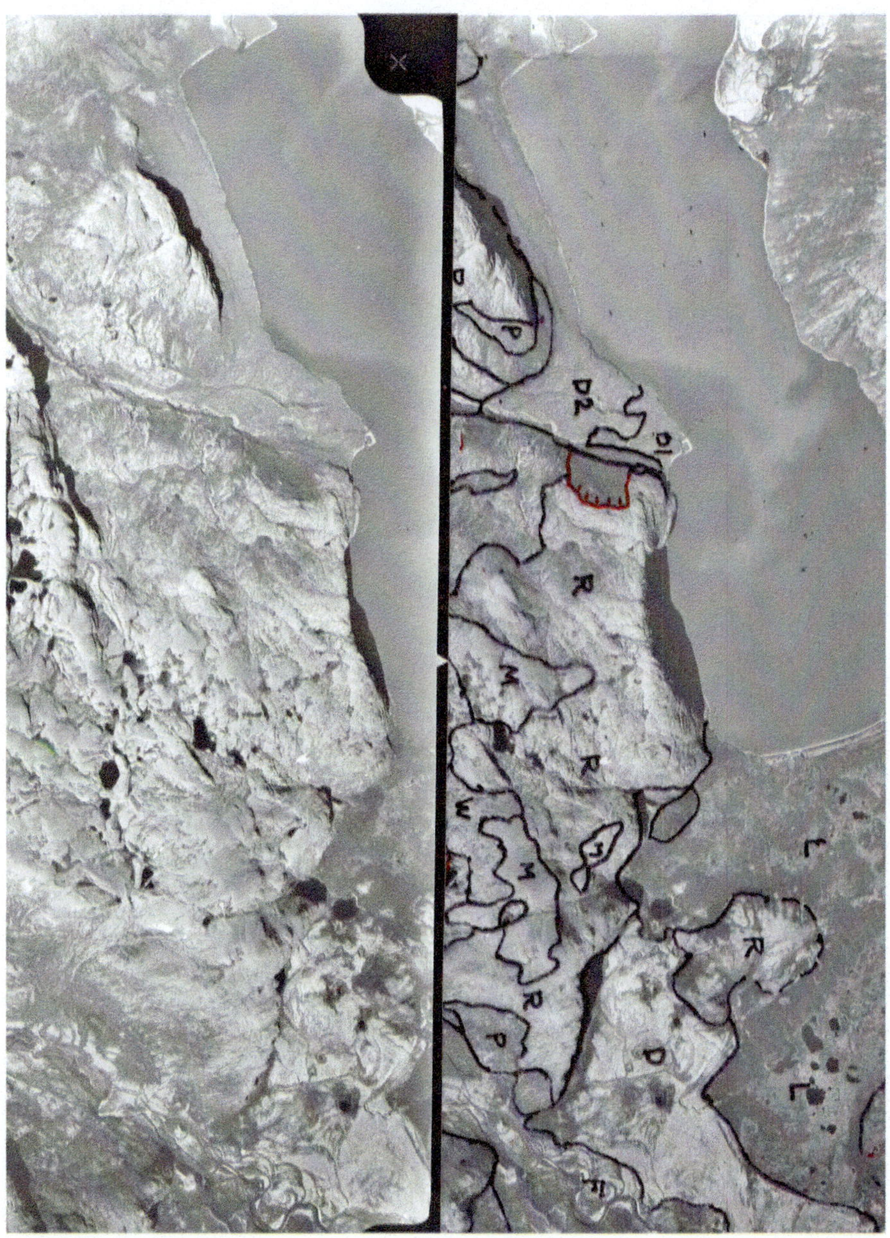

G15 Stereopair

Photo Interpretation

The 1:60,000 scale photo covers 117 km^2. Eleven units in five categories occur in the site.

Glacial Deposits

M—Hummocky moraine covers the bedrock at 500 m altitude in the center of the photo. This moraine is in a zone of thicker till than is usual for north Labrador.

W—an area of reworked moraine slopes down toward the delta stream from 375 to 220 m elevation

Glaciofluvial Deposits

X—spillway channels drain through all the lake-filled depressions in the bedrock upland

Red lines—trace a number of east-west oriented eskers that are associated with the spillways

D—Two deltas occur at and near Umiakovik Lake. The larger bare-appearing unit occurs at the lacustrine extension of the lake at 40–60 m elevation. The smaller forested unit debouches into the lake in three levels. **D1** lies barely above the lake level. **D2** is at 40 m, while **D3** occurs along the stream gorge at 145 m.

Alluvial Deposits

F—a partly wooded braided stream flows along the east side of the site

f—a small alluvial fan at the base of a plunging stream lies on the west side of the braided stream

Glaciolacustrine Deposits

L—fine and wetland sediments fill the eastward extension of Lake Umiakovik

Bedrock Units

R—all bedrock in the site is Mid Proterozoic granite

Photo Source

Courtesy of National Air Photo Library, A18299, 60-61

Select Bibliography of Canadian Glacial and Alpine Glacial Terrains

Clark, P. U. (1984). Glacial geology of Kangalaksiorvik-Abloviak region, northern Labrador, Canada; unpublished PhD thesis, University of Colorado, Boulder.

Ives, J. O. (1959). Glacial geomorphology of the Torngat Mountains, Northern Labrador. *Geographical Bulletin 12*.

Klassen, R. A., Paradis, S., Bolduc, A. M., & Thomas, R. D. (1992). Glacial landforms and deposits, Labrador, Newfoundland and eastern Québec. *Geological Survey of Canada*, Map 1814A, scale 1:1,000,000.

Lopoukhine, N., Prout, N. A., & Hirvonen, H. E. (1977). Ecological land classification of Labrador, Lands Directorate (Atlantic Region) Environmental Management Services, Fisheries and Environment Canada, Halifax, Nova Scotia. *Ecological Land Classification Series, No 4*.

Ryan, B. (compiler) (1990). Preliminary geological map of the Nain Plutonic Suite and surrounding rocks (Nain-Nutak, NTS 14 S.W.), scale 1:500,000, Geological Survey Branch, Department of Mines and Energy, St. John's Newfoundland, Map 90–44.

Ryder, J. M. (1998). Geomorphological processes in the Alpine areas of Canada. *Geological Survey of Canada Bulletin 524*.

Vernon, P., & Hughes, O. L. (1966). Surficial geology, Dawson Creek, Larsen Creek, and Nash Creek map areas. *Geological Survey of Canada Bulletin 136*.

Air and Ground Photo Collections for the European Alpine Terrains

Alpen Flugbild (undated). 62 Flugaufnahmen von Werner Friedli, Weltflugbild Verlag, Feldmeilen, Schweiz.

Das Panorama der Alpen Mit 30 Panorama-Fotos, (*triple page spreads*) Rudolf Rother, Christof Stiebler, Franz Thorbecke, Suddeutscher Verlag Munchen, 1969.

A direct terrain-dependent, extensively exploited prime economic resource of the European Alps is given in a geotechnical inventory *Speicherseen der Alpen,* Sonderheft Wasser-und Energiewirtschaft 9/1970.

Periglacial Terrains

Periglacial terrains occur in seasonally unfrozen unconsolidated deposits and organic materials in high latitude and high altitude environments. The Examples are taken from the three ordered Groups:

***Zi*—Ground Ice; *Zm*—Cryoturbated materials; *Zk*—Thermokarst terrain**

The definition and description of periglacial terrains can be found in the *Glossary of Permafrost and Related Ground ice Terms,* National Research Council of Canada, Technical Memorandum 142, 1988

Ground Ice Units

Figure Zi1: Pingos

Location 69°24′ 30″N, 133°05′23″W

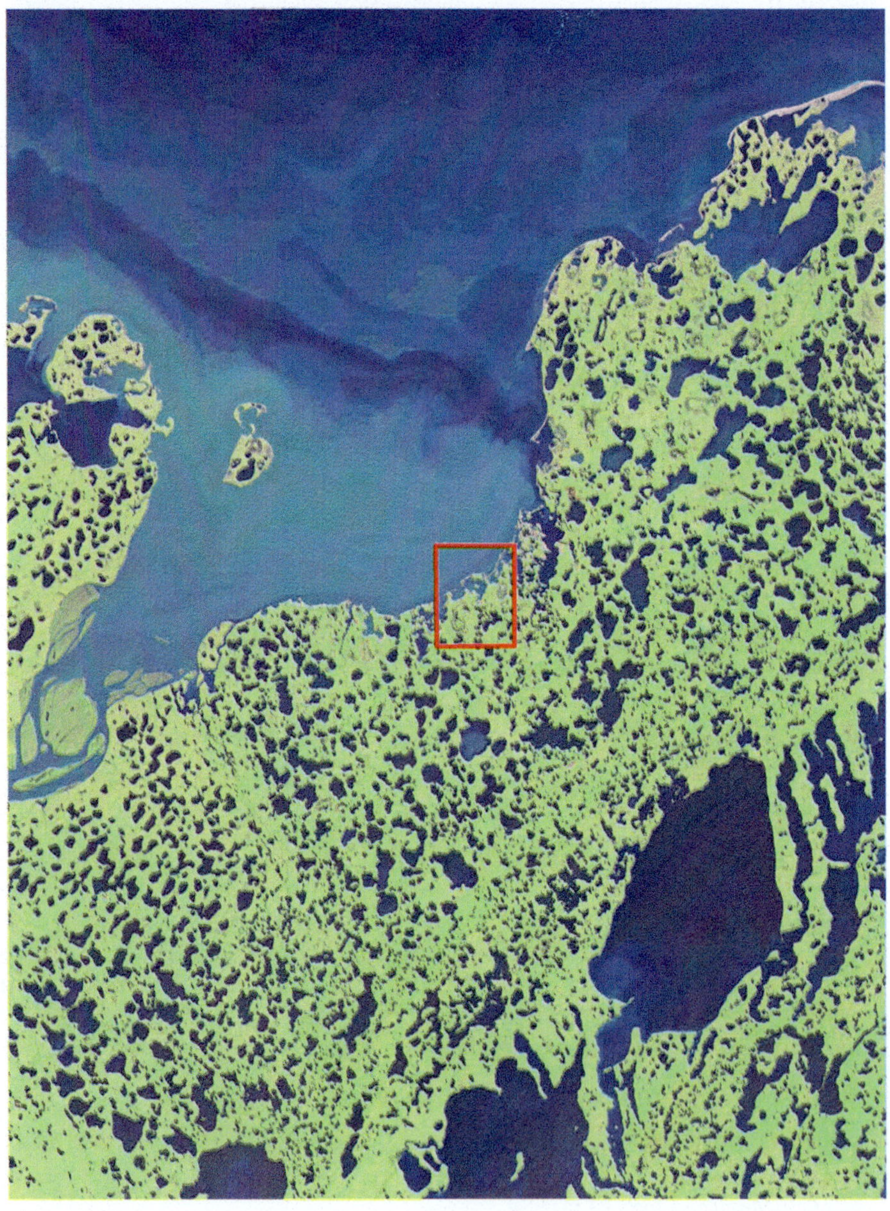

Landsat Image T/M 4, 10 July 2013

Regional Environment

The image covers an area of 6,620 km^2 east of the Mackenzie Delta on the Arctic Coastal Plain. The myriad lakes are thermokarst in glaciomarine and lacustrine sediments of the Arctic Coastal Plain. The Example site is on Kugmallit Bay of the Beaufort Sea.

Finger-like ridges on the east edge of the scene are ground moraine etched by meltwater valleys.

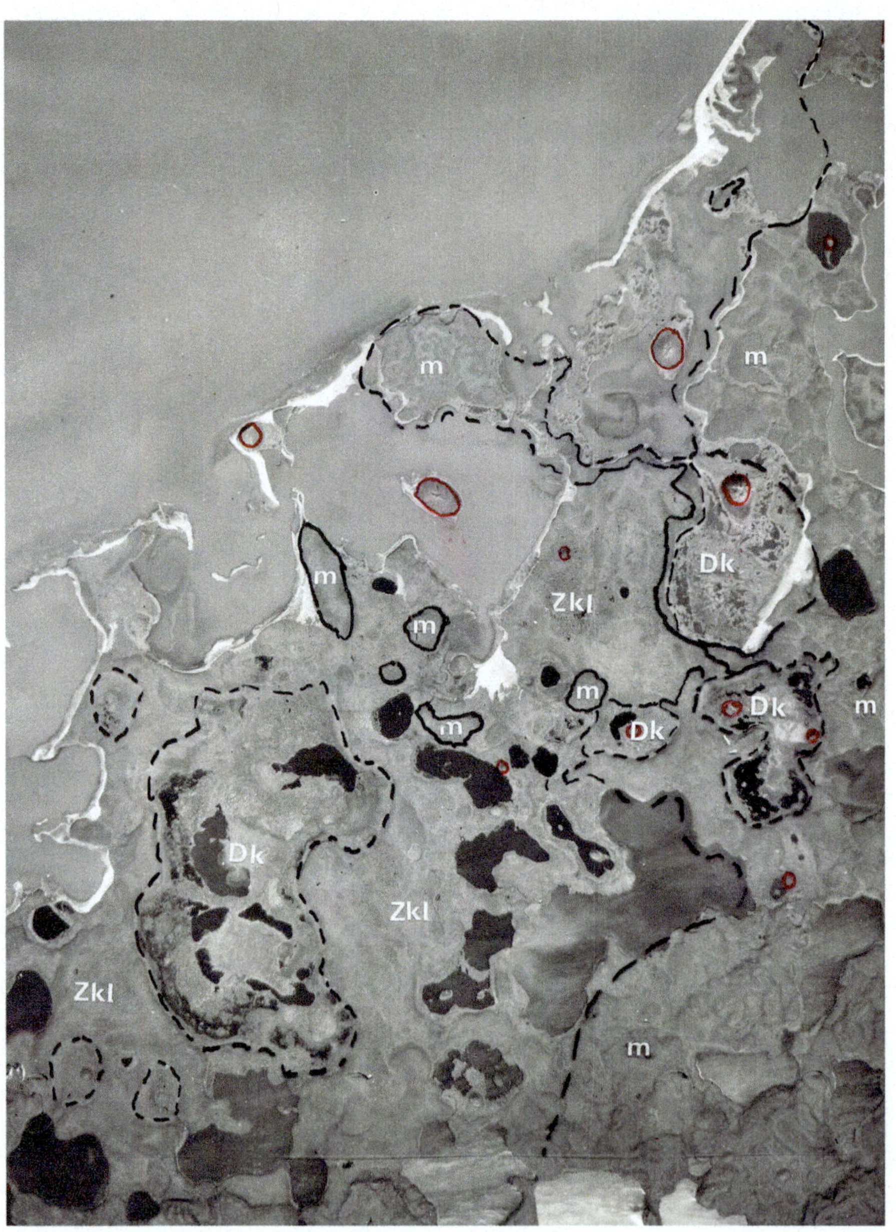

Zi1 Airphoto

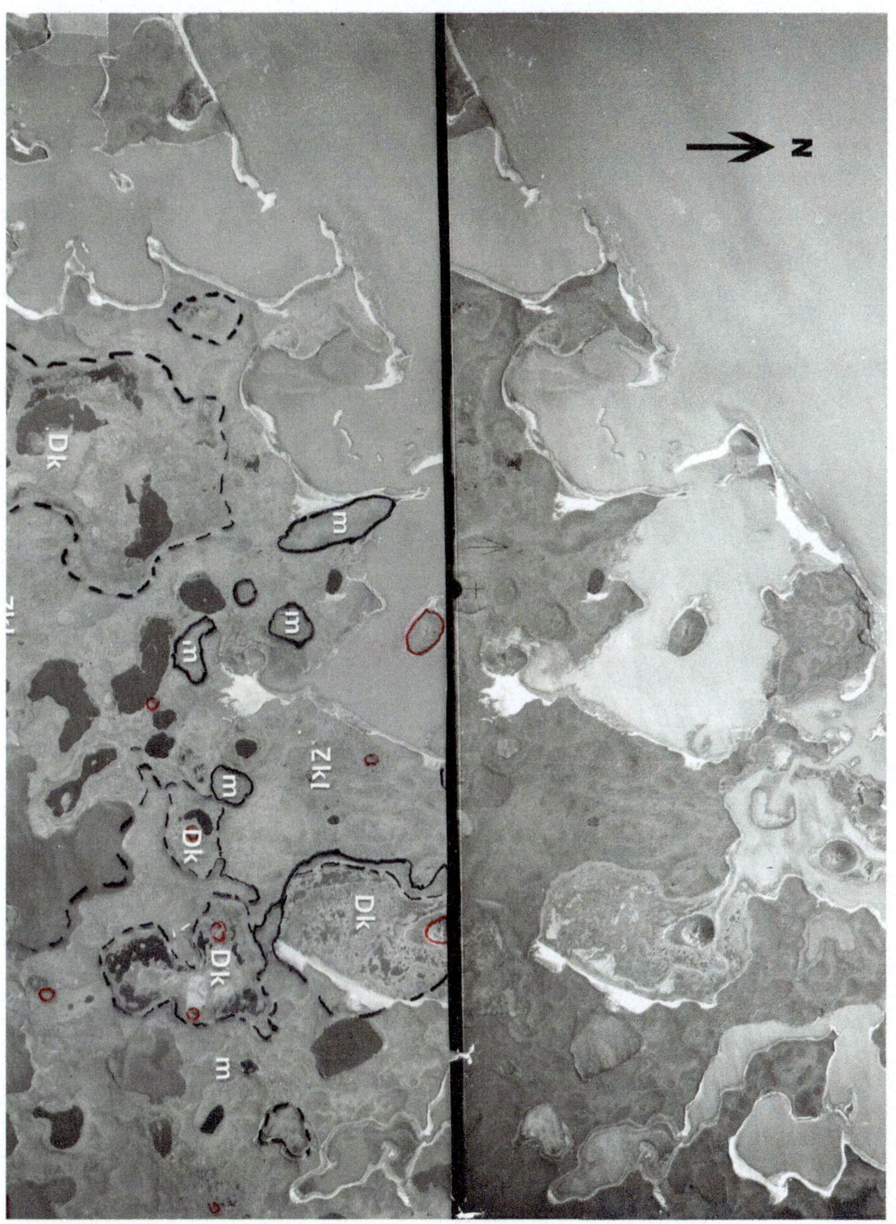

Zi1 Stereopair

Photo Interpretation

The 1:37,000 scale air photos locate four large pingos, circled in red, south of the port of Tuktoyaktuk. There are 17 smaller pingos detectable in the photo area, and 1,400 in the Tuktoyaktuk area. They developed between 10 and 1 ka but many are less than 2,000 years old. (6,000 similar pingos have been identified in northern Asia).

The greater part of the site is occupied by lacustrine sediments which filled large coalesced **Zk1** thermokarst basins (Rampton 1988) and consist of clay, silt and fine sand 2–8 m thick with peat capping. The active layer is 0.5 m.

M are glacial till deposits. Areas coded **Dk** are lake basins which initially developed through thermokarst subsidence and were subsequently drained.

Photo Source

Courtesy of National Air Photo Library, A12918, 93-94

Figure Zi4: Ice-Wedge Polygons

Location 71°30′N, 79°22′W

Landsat Image T/M 4-5, 14 August 2005

Regional Environment

The Example area is on northern Baffin Island of the northeast Canadian Shield.

Structure

The scene is bisected by a distinct northwest to southeast lineament marked by a string of lakes. It is the east end of the 105 km long Phillips Creek Fault Zone that is seen on Figure **G2** 40 km to the northwest. The lake-studded area on the south is the east end of the Ordovician limestones of the Phillips Creek Trough. The area on the north is Archean granite gneiss.

The site is 18 km north of the marked Mary River iron ore development area.

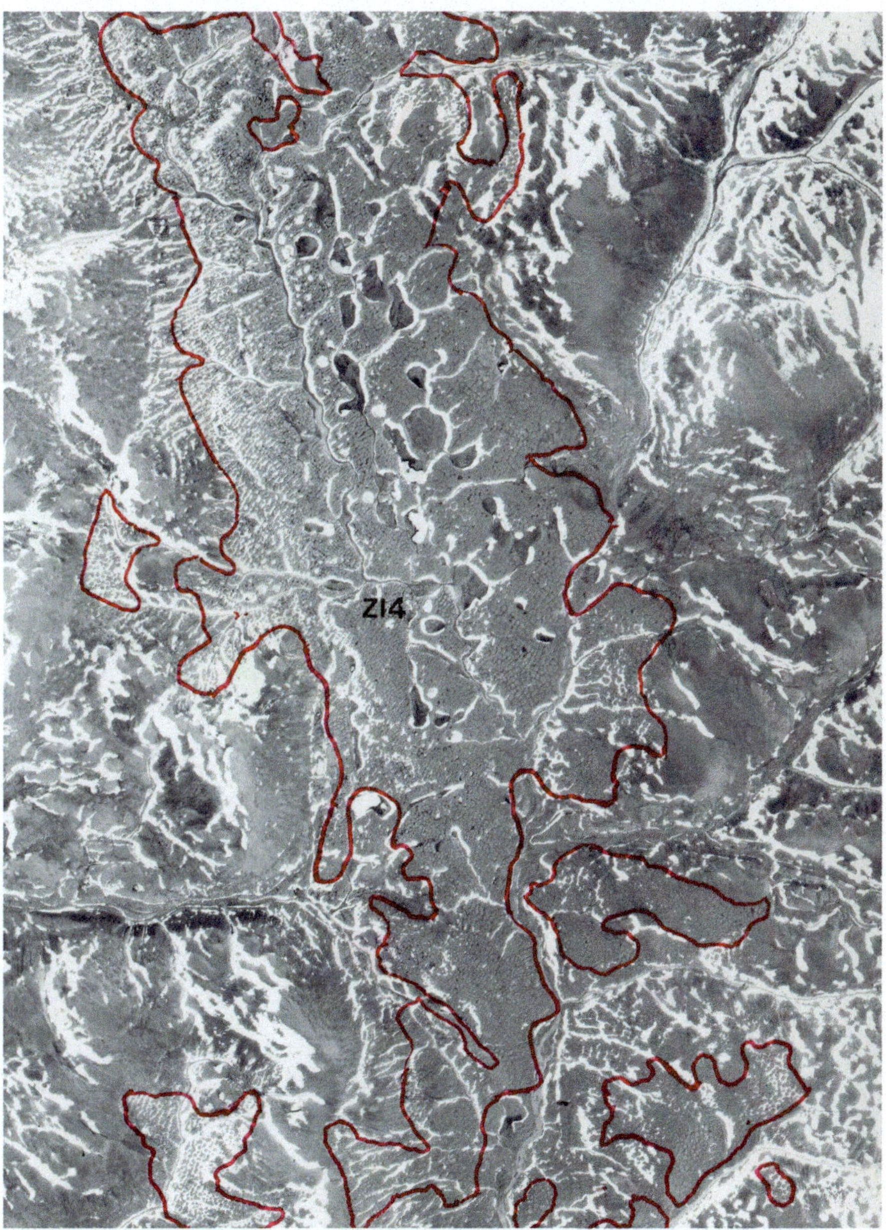

Zi4 Airphoto

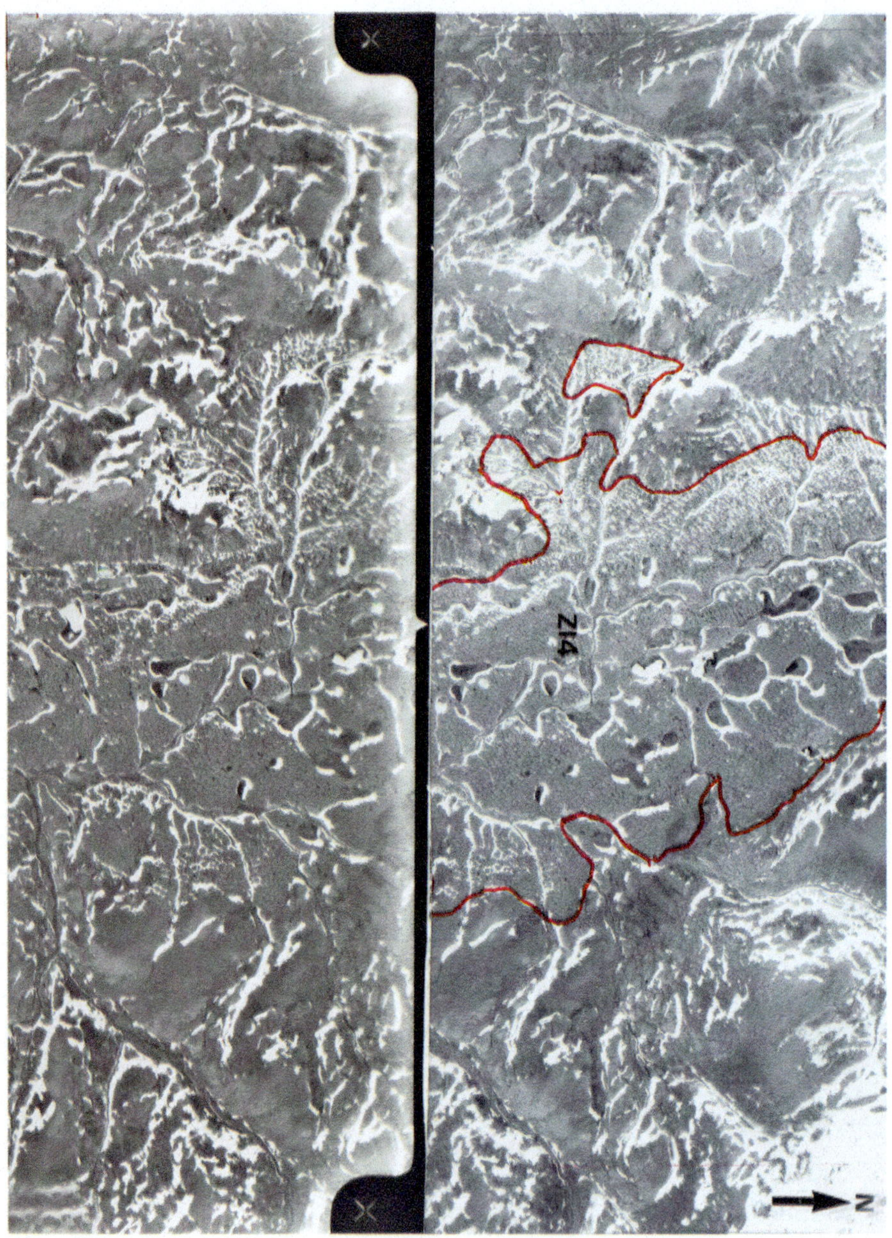

Zi4 Stereopair

Photo Interpretation

The 1:60,000 photos taken on 3 July 1958 delineate a large concentration of high center polygons in a broad shallow, probably fault-related valley of deposits of 2 m thick glacial till with an approximately 20 cm active layer. The side valley is at 475 m elevation while the adjacent hills range from 675 to 725 m.

Photo Source

Courtesy of National Air Photo Library, A16107, 160-161

Cryoturbated Materials

Figure Zm1.2: Gelifluction Stripes

Location 67°55′N, 116°30′W

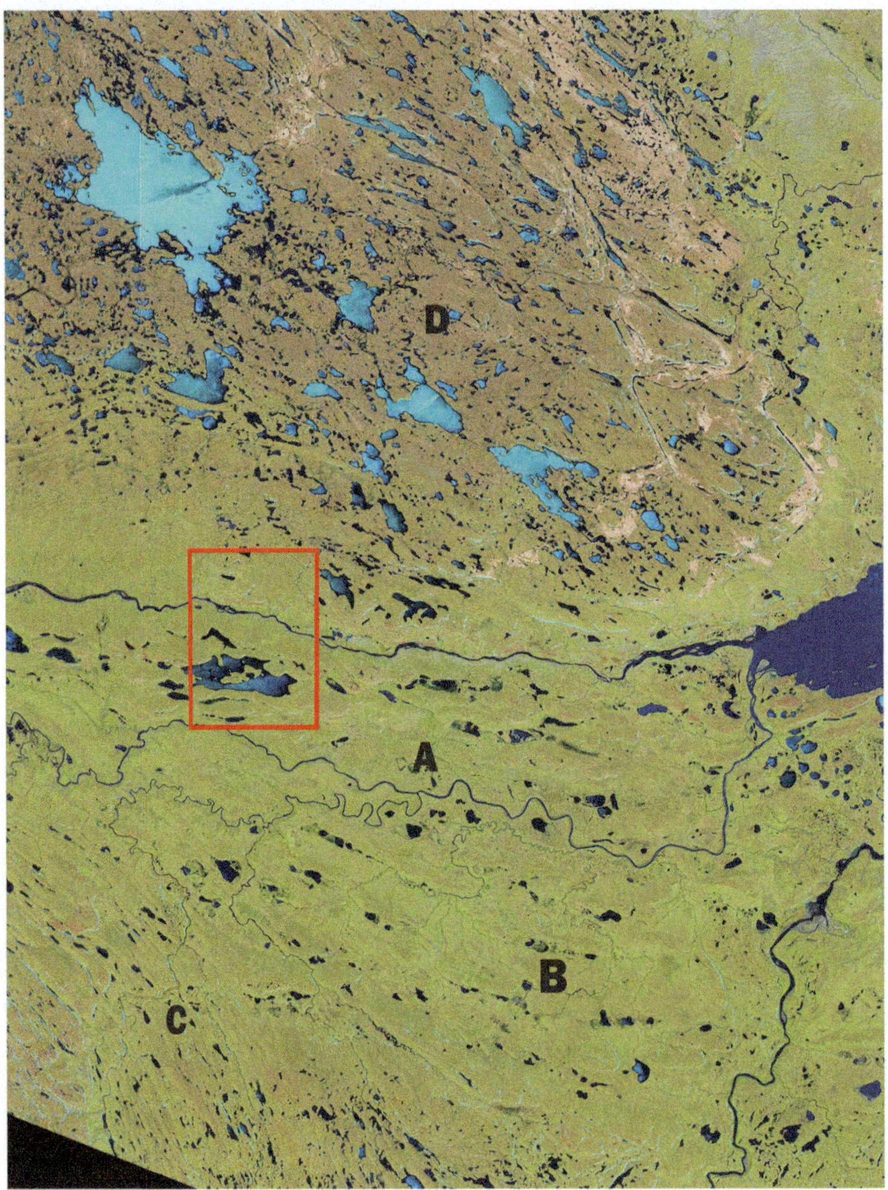

Landsat Image T/M 4-5, 01 July 2011

Regional Environment

The scene covers 7,000 km^2 on the Arctic Platform of the Northern Interior Plains. The Rae River that crosses through the photo area bisects the scene into two distinct geological terrains. It and the parallel Richardson River flow to Coronation Gulf on the right.

Zones **A, B,** and **C** are in the extreme northwest of the Canadian Shield.

- Zone **A** is a 20 km wide belt of Proterozoic argillite intruded by gently north dipping gabbro sills
- Zone **B** are glaciomarine deposits covering the argillite rocks
- Zone **C** are northwest trending ice flow flutings in glacial till
- Zone **D** The brown lake-studded area consists of drumlin fields and ground moraine deposits overlying Paleozoic carbonate sediments.

Zm1.2 Airphoto

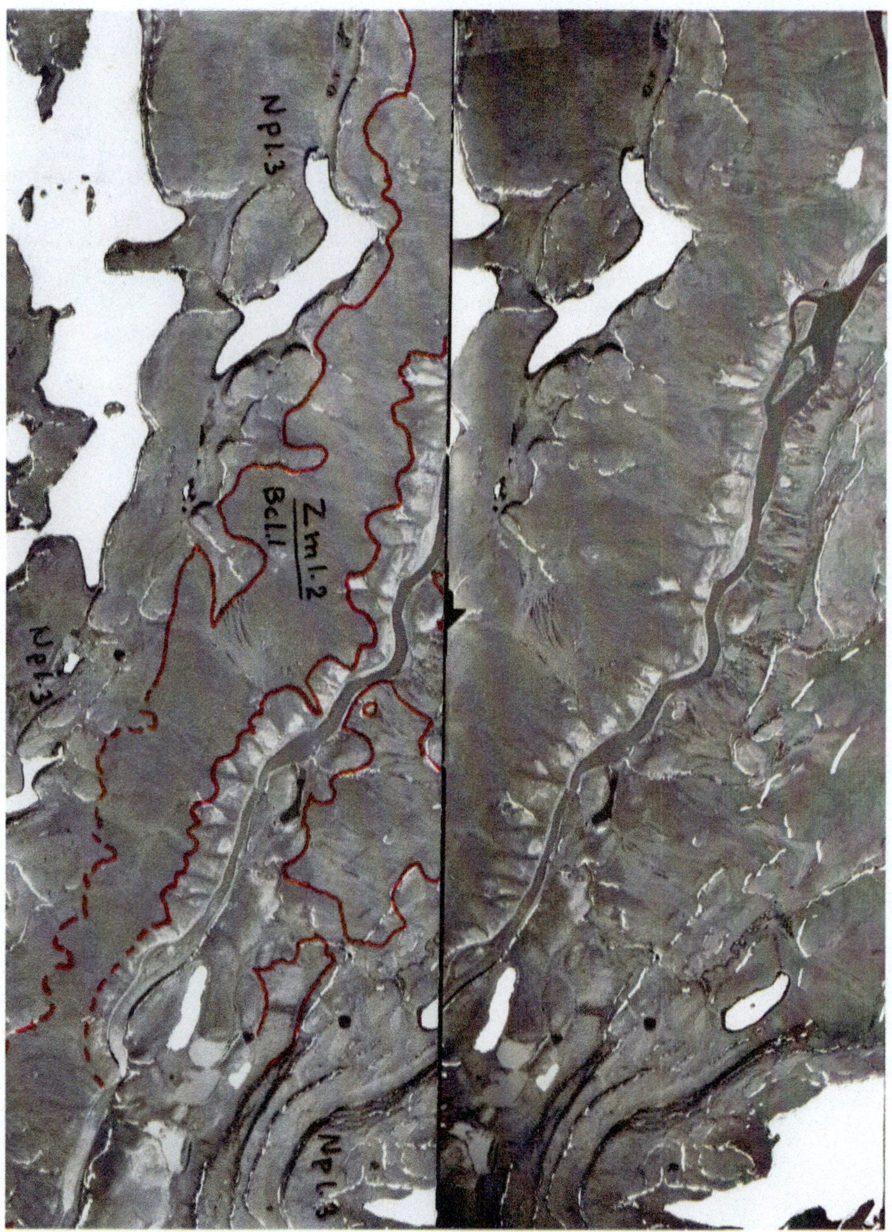

Zm1.2 Stereopair

Photo Interpretation

The 1:60,000 air photo covers an area of 120 km^2. It shows two extended deposits of gelifluction stripes.

The north unit is at 170 m elevation, descending on deposits of glacial till coded **Gt** to the Rae River at 50 m elevation.

The south unit extends along the river valley. The stripes flow on marine deposits down the dip of the sills to the river bank.

Areas coded **Np1.3** are elongated bedrock hills with 40–50 m cliff-like southern faces dipping gently northward. They consist of 720 ma gabbro sills that were intruded into 1,900 Ma sedimentary rocks.

Photo Source

Courtesy of National Air Photo Library, A13608, 12122

Figure Unit Zm2: Rock Glaciers

Location 45° 20′30″N, 06°39′45″E

Landsat Image T/M 4-5, 16 October 2003

Regional Environment

The site is in the Carboniferous sedimentary zone of the Vanoise National Park of the French High Alps. Blue areas are snow cover not ice.

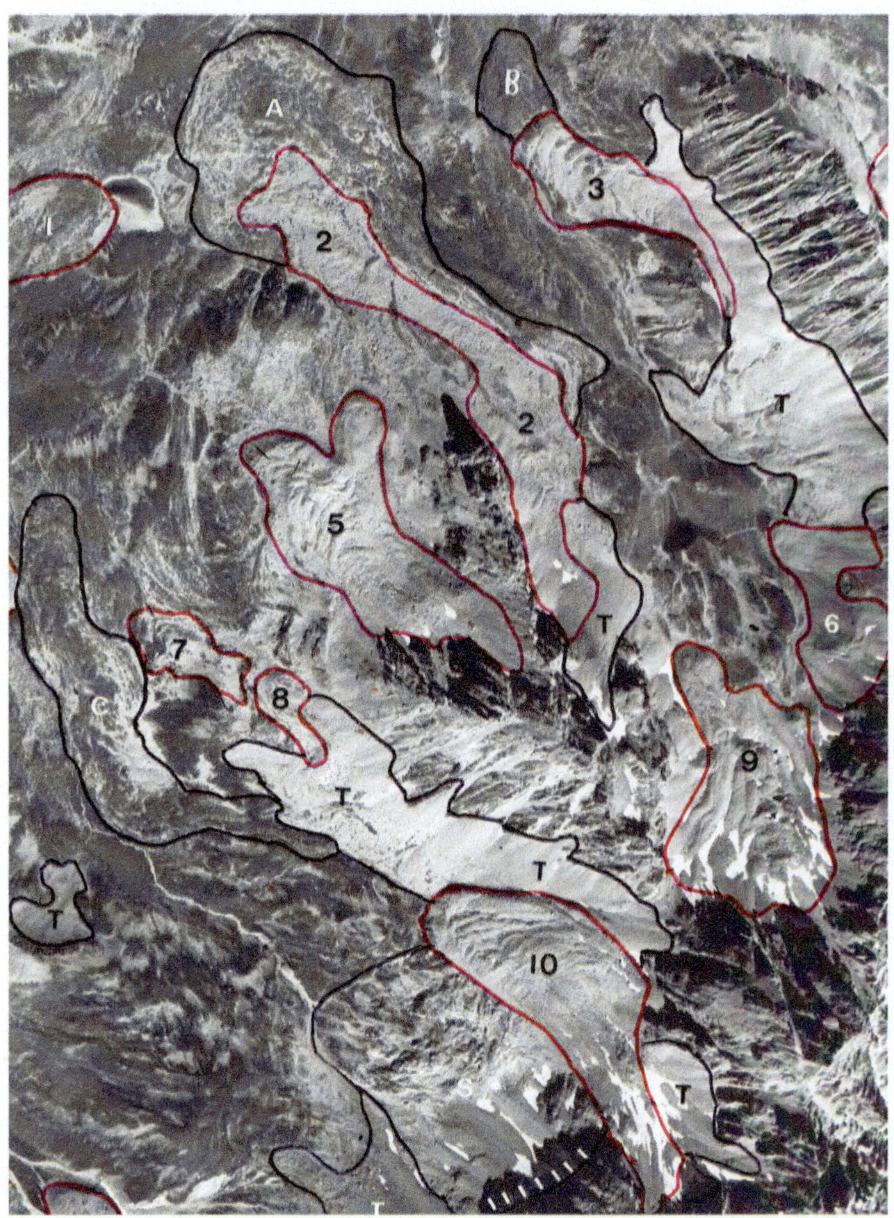

Zm2 Airphoto

Photo Interpretation

The site of this **single** 1:14,000 scale air photo enlargement is at 2,500 m elevation. Eleven numbered *active* and three coded **A, B, C** *inactive* rock glaciers are delineated (# 4 off photo frame). The inactive glaciers are vegetated. The white, smooth-appearing units coded **T** are active talus slopes.

The unit coded **S** is a rock slide with the plane of the slip surface and the landslide block mass clearly distinct.

Photo Source

Figure Zm3: Detachment Failures

Location 79°55′40″N, 85°58′39″W

Landsat 7 Image S2C, 08 July 1999

Regional Environment

The site is located in Jurassic sediments on the south side of Slidre Fjord on the Fosheim Peninsula of northern Ellesmere Island.

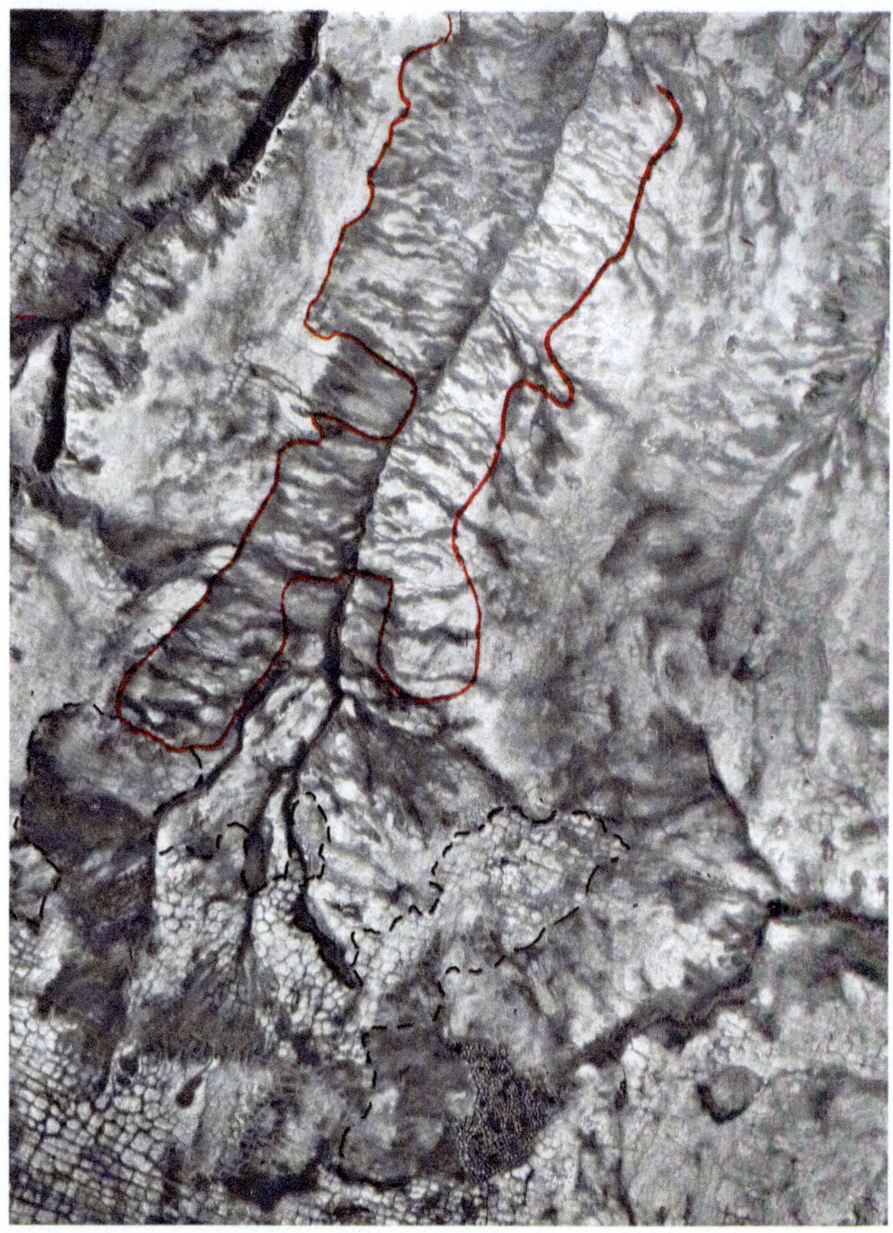

Zm3 Airphoto

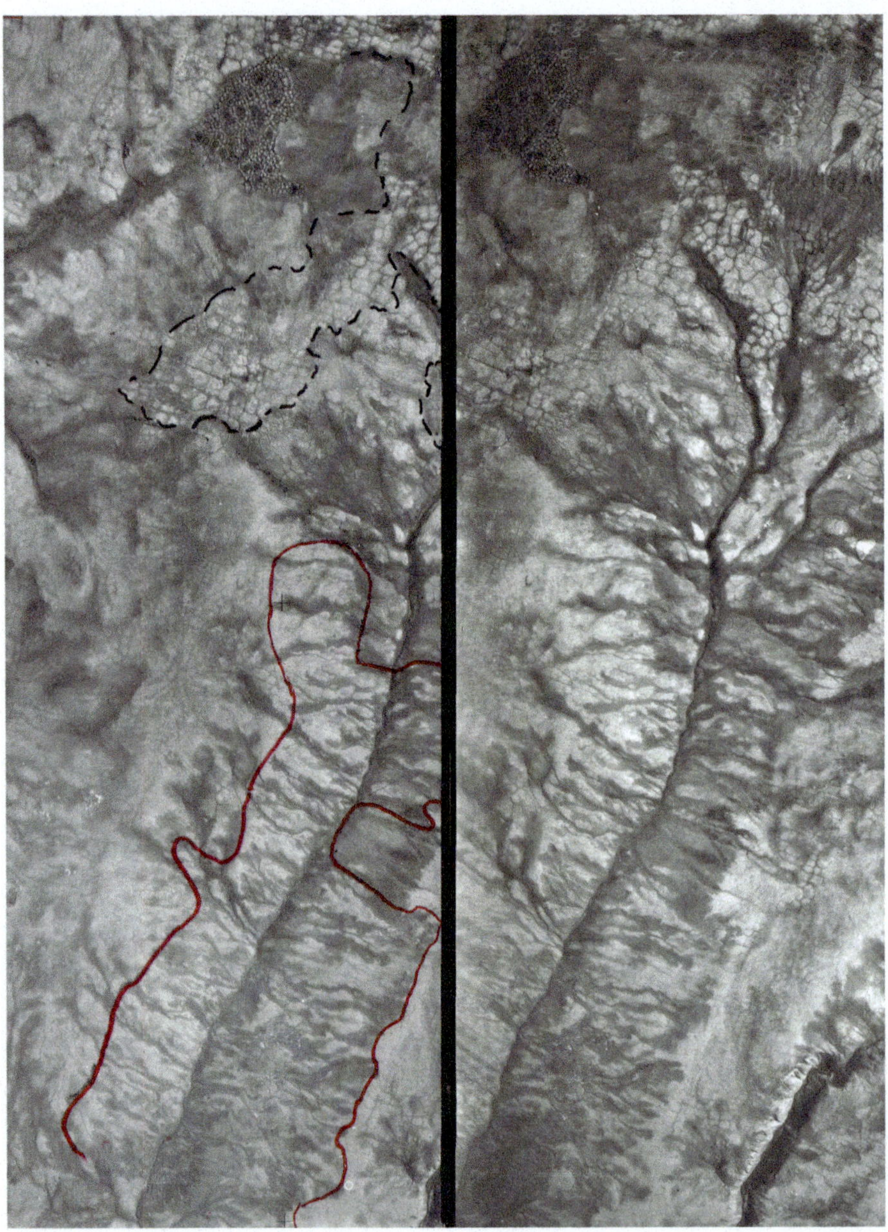

Zm3 Stereopair

Photo Interpretation

The 1:18,000 scale photos show a delineated zone of unusual concentration of dark, moist erosion stripes 30–40 m wide and 300–500 m long at 35 m elevation. The stripes are tentatively interpreted as detachment failures delineated on either slope of a narrow stream valley that flows down to the fjord in thick till. Active layer depths reportedly range from 40 to 90 cm.

Such concentration is related to the fact that the region is anomalously warm for its latitude in summer due to surrounding mountain ranges which tend to block or dissipate slow-moving disturbances from the Arctic Ocean and from Baffin Bay.

Large concentrations of ice wedge polygons are prominent in the south half of the photos.

Photo Source

Courtesy of National Air Photo Library, A12725, 223-224

Reference

Duguay C. R., Lewkowicz A., (1995). Assessment of SPOT panchromatic imagery in the detection and identification of permafrost features Fosheim Peninsula, Ellesmere Island N.W.T., 17th Canadian Symposium on Remote Sensing 1995, pp 8–15.

Thermokarst Terrain

Figure Zk1: Subsidence Terrains

Location 69°05″N, 137°50′W

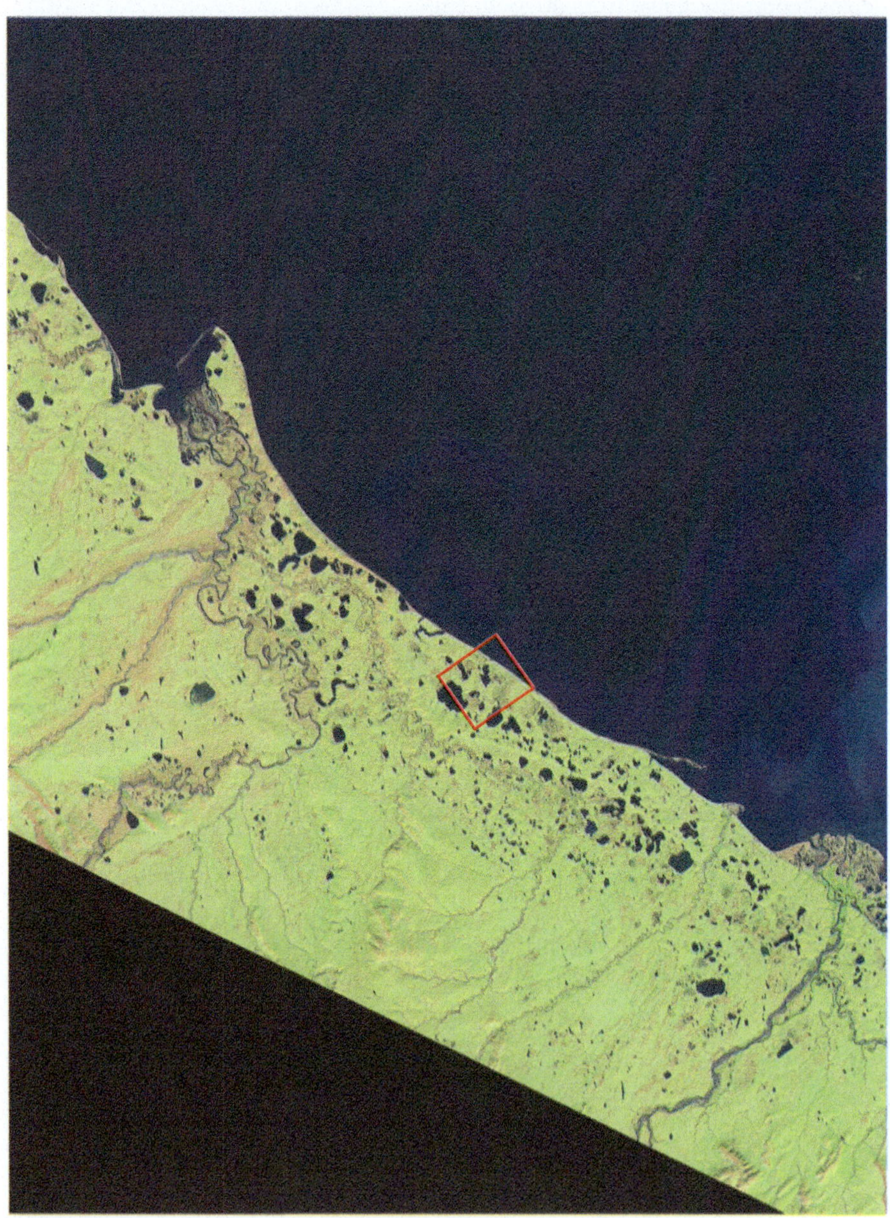

Landsat Image T/M 4-5, 06 September 2009

Regional Environment

The site is on the Yukon Coastal Plain of the Beaufort Sea 50 km west of the Mackenzie Delta.

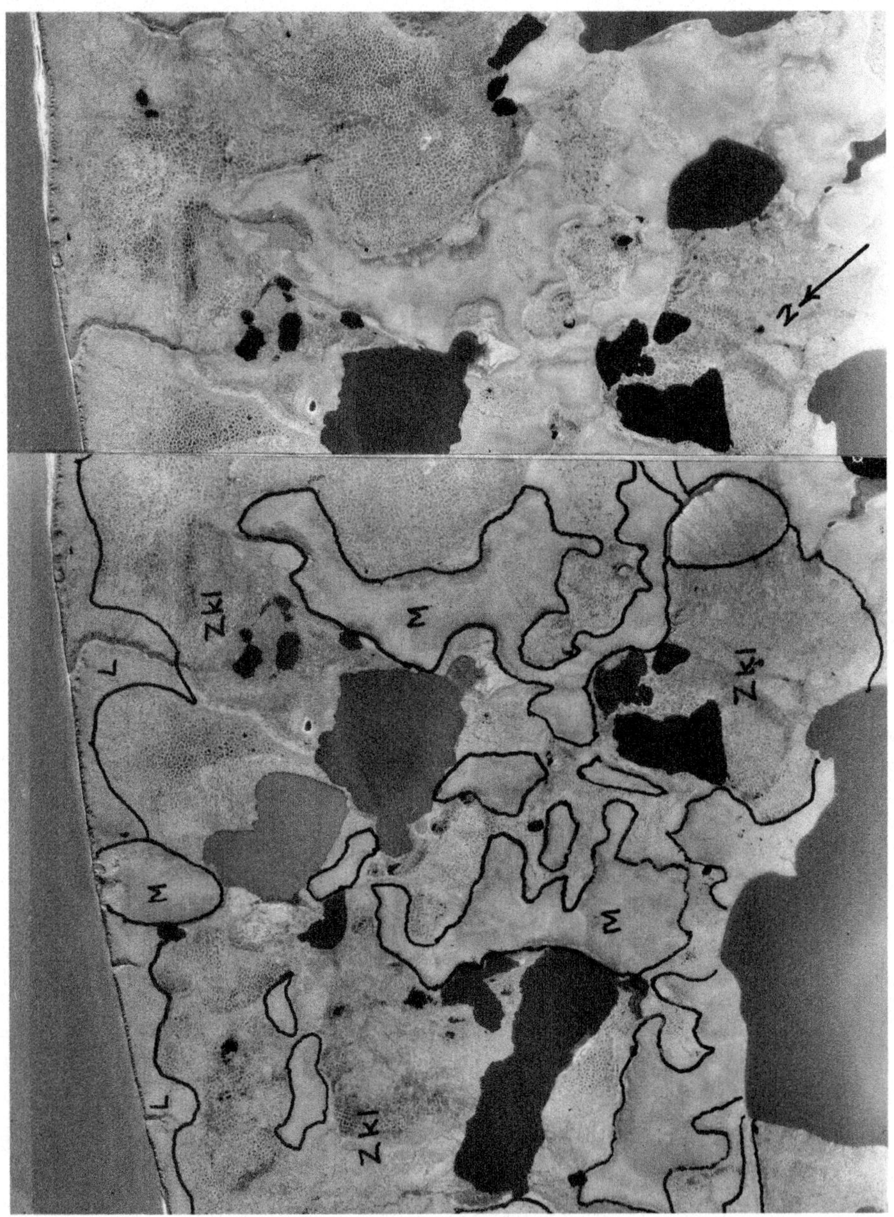

Zk1 Airphoto

Photo Interpretation

This 1:33,000 **stereogram** is located on glaciolacustrine sediments 3–9 m thick. **Thaw lakes** are underlain by glacial till. **Zi4** ice wedge polygons cover all the lacustrine sediments. Pond waters act as heat sinks melt the ice beds underlying them and expand through slumping and subsidence creating a thaw lake. The sediments consist of silty clay and silty sand with layers of peat. They are underlain and surrounded by outcrops of **M** coded till deposits. **L** is the exposed coastal bluff of the local 5 km wide lacustrine plain.

Photo Source

Courtesy of National Air Photo Library, A26779, 11-12

References

Frohn, R. C., et al. (2005). Satellite remote sensing classification of thaw lakes and drained thaw lake basins on the North Slope of Alaska. *Remote Sensing of Environment, 97*, 116–126.

Rampton, V. N. (1988). Quaternary geology of the Tuktoyaktuk Coastlands. *Northwest Territories Memoir 423*.

Non-glacial Surficial Terrains

This category of complex terrains contains an Example of each of four types of surficial deposits: fluvial; marine; lacustrine; and aeolian from Canada, USA, and France.

Fluvial Deposits

Figure S1: Tuchan

Location 42°52′N, E02°43′E

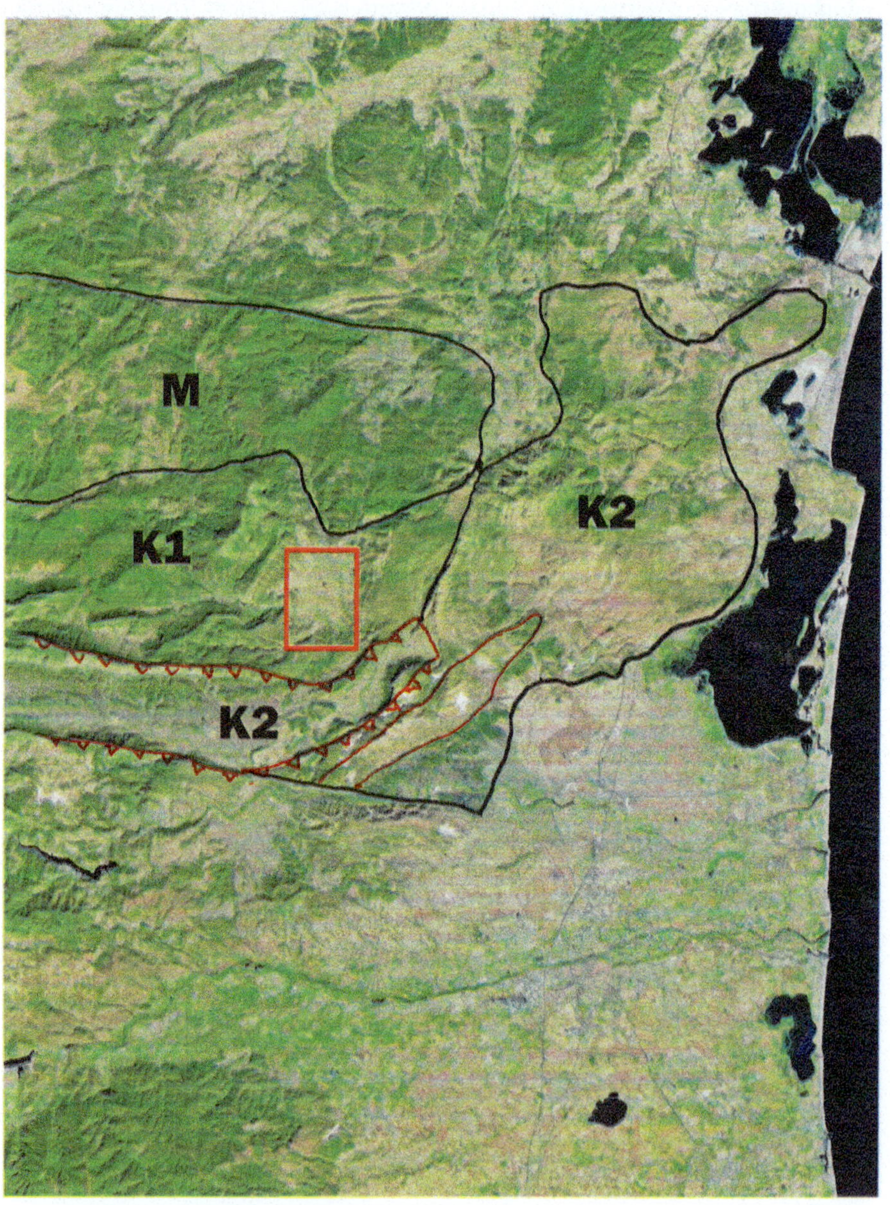

Landsat Image T/M 4-5, 04 October 2011

Regional Environment

The scene covers 1,700 km^2 on the Mediterranean coast of southwest France.

Structure

The area consists of a complex zone of thrusted and folded Mesozoic sedimentary rocks in the center flanked by folded and thrusted Tertiary sediments on the north and Quaternary marine sediments of the coastal plain of Roussillon on the south. Saline lagoons behind barrier beaches line the coast.

The complex terrain in which the Example is located is in the east-west longitudinal North Pyrenean tectonic Zone, divided into an anomalous district and a group of three related districts that make up the Corbières Range. The range is a nappe thrust sheet moved northward on a base of Triassic marl.

District **M,** mainly forested, is a rugged upland outlier of the Central Massif. It is made up of Mid Paleozoic interbedded sedimentary rocks that have been folded and faulted by Hercynian and Pyrenean tectonics.

District **K1** is a folded upland 200–600 m elevation of Triassic, Jurassic and Upper Cretaceous limestones

District **K2** is a synclinal lowland of 130 and 180 m elevation of Lower Cretaceous limestone between two topographically distinct thrust faults.

District **K2** is the eastern part, 70–500 m area of folded Lower Cretaceous limestones.

A distinct northeast-striking bare anticlinal ridge is traced near the thrust fault.

Physiography

The example area is a 15 km^2 intermont basin 25 km inland in the **K1** structural district. It is enclosed by Mesozoic calcareous rocks rising to 600 m elevation on the west and 250–300 m on the east. The basin is infilled with Pliocene alluvial sandstones and marls of the Tarrasac and Verdouble Rivers. The rivers have cut a number of water gaps through the northern thrust ridge to flow into the **K2** lowland

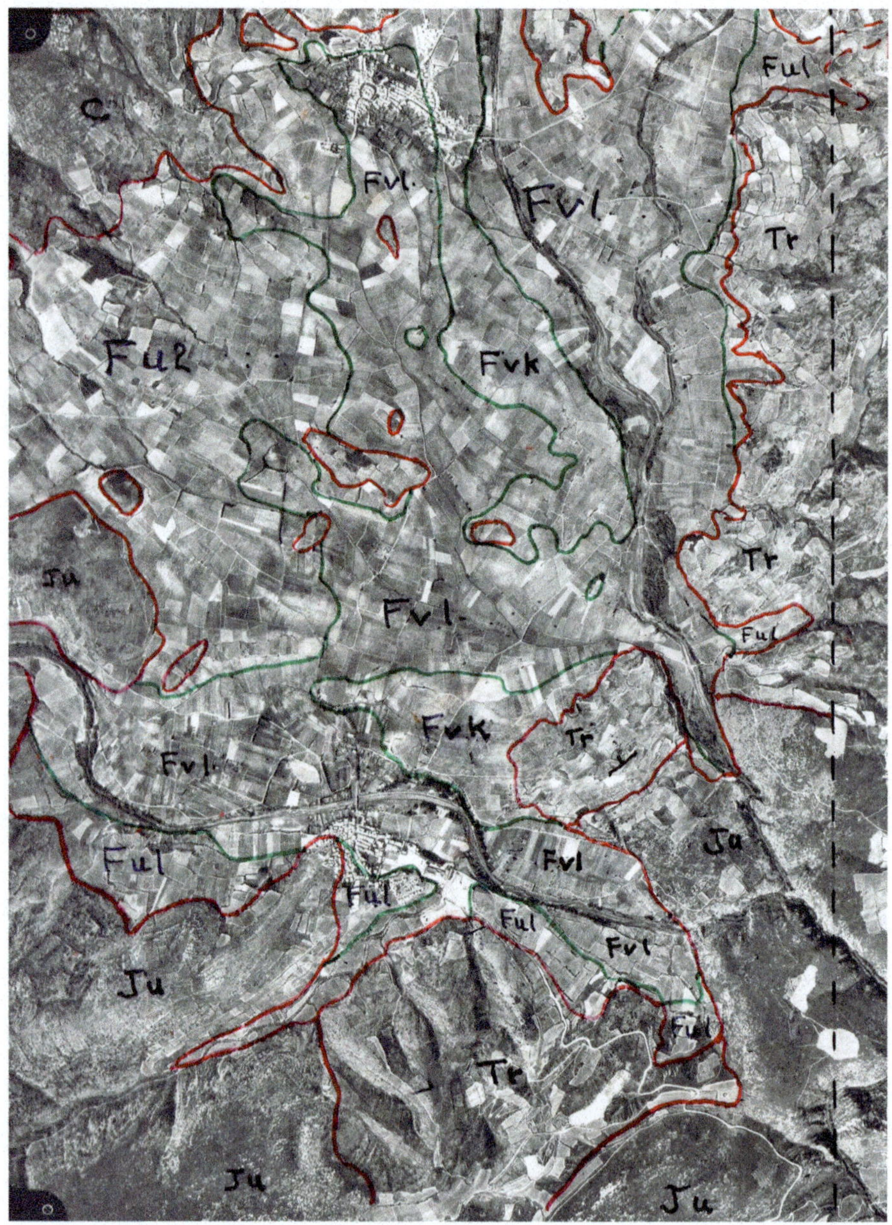

Ful
C
Fvl.
Fvl
Tr
Ful
Fvk
Tr
Ju
Fvl
Ful
Fvk
Tr
Fvl
Ju
Ful
Fvl
Ful
Ful
Fvl
Ju
Ful
Tr
Ju
Ju

S1 Airphoto

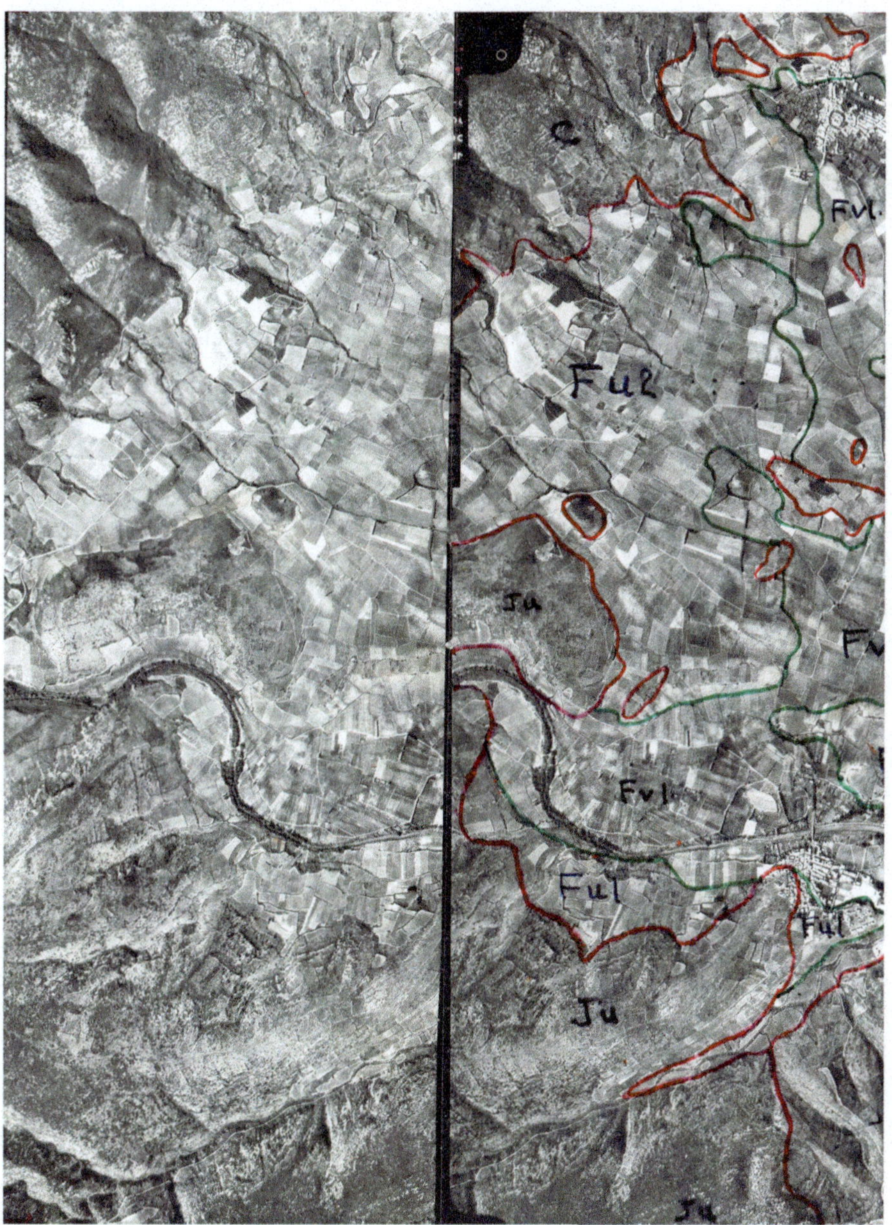

S1 Stereopair

Photo Interpretation

The 1:25,000 scale air photo area is 22 km^2. There are occurrences of four units of fluvial deposits, and three units of surrounding Mesozoic sedimentary rocks. Viticulture is the single land use of all the surficial units

Fluvial Units

Fv1—This is the most extensive and youngest unit, occupying the low terraces of the two river valleys, ranging from 130 to 140 m in elevation. The sediments are stratified Holocene gravel and/or sand with a minor fraction of silt.

Fu1—these-are alluvial fans bordering the river terraces

Fvk—these are residual sandstone and marl terraces at 150 m elevation of Pliocene river deposits

Fu2—this is a 2 km broad and 2 km long Pliocene alluvial fan sloping into the basin from its apex at 235 m at the foot of the western rock hills to its toe at 150 m

Bedrock Units

Tr—these are dissected hills of interbedded Triassic sediments at 160–200 m elevation

Ju—these are massive-appearing scrub-covered hills of Jurassic calcareous rocks rising from 200 to 250 m

C—this a lower slope of Cretaceous calcareous rocks at 335 m elevation

Photo Source

Copyright IGN, 1976, 2810, 1581-1582

Figure S2: Burlington

Location 44°30′N, 73°13′W

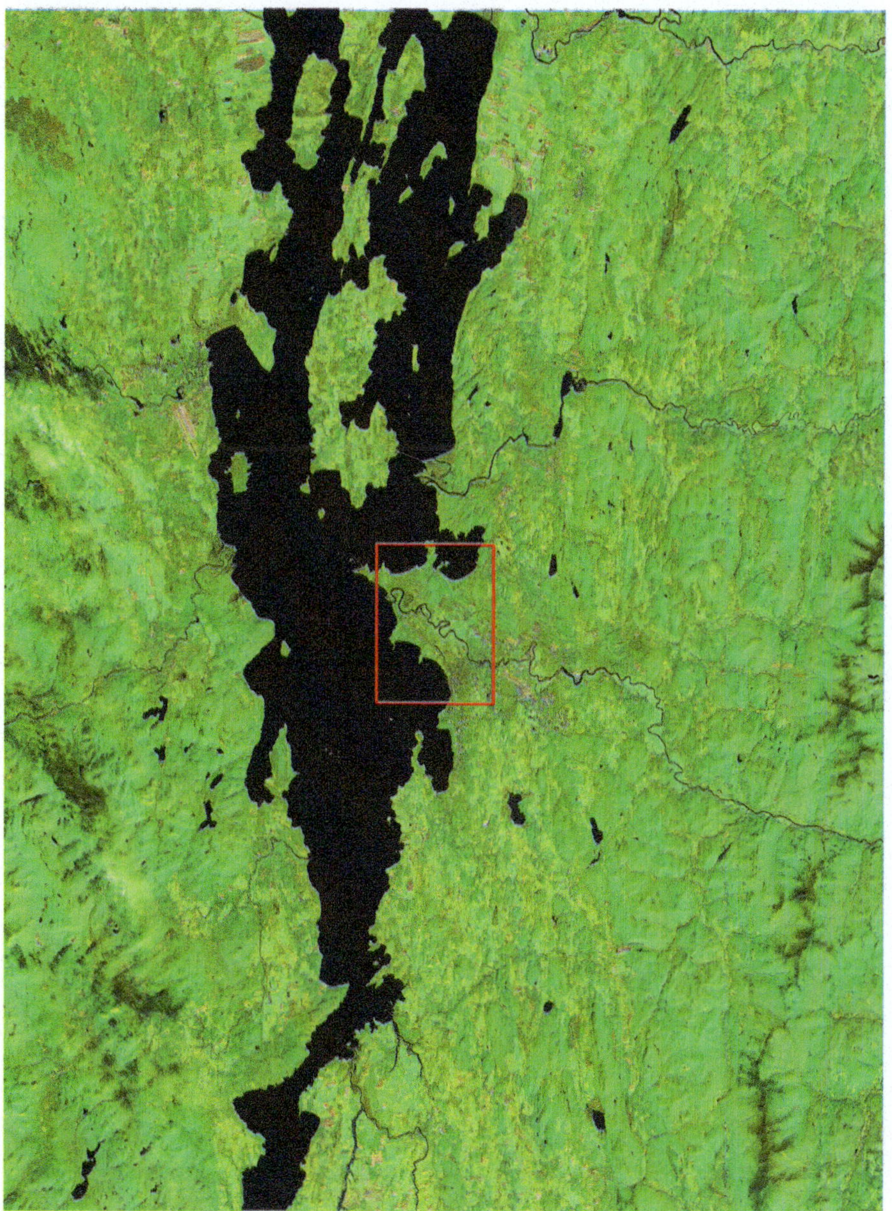

Landsat Image T/M 4-5, 25 September 2009

Regional Environment

The scene shows the Example as a delta jutting into Lake Champlain in the state of Vermont, USA, 50 km south of the Canadian border.

Structure

The site area is located in the Champlain Lowland, a structural trough between the Green Mountains and the Adirondack shield rocks of New York State on the west. The trough consists of folded Lower Paleozoic rocks broken by north-south striking east-dipping thrust faults which have no particular topographic expression. Wisconsin glaciations deposited extensive layers of ground moraine in the lowland. These tills were later covered by the post-glacial invasion of marine waters from the St. Lawrence Champlain Sea which deposited the local 3 m thick sands.

S2 Airphoto

S2 Stereopair

Photo Interpretation

The 1:60,000 scale natural colour air photo covers 120 km^2 in which occurrences of seven geomorphic units are delineated.

Bc1—Three meters of marine sands and gravels overlying silts and clays occur as a delta. The delta was formed by the Winooski River and its local tributaries eroding and redepositing marine deposits lying inland at elevations ranging from 40 to 100 m.

Fv2—This is the 3 km broad 30 m elevation floodplain of the Winooski River. The fine sands and silts 1–7 m thick of the meandering channel have carved into the easily erodible **Bc1**delta marine sands.

Fv2T—this is a 7 m high river terrace

Y2—these are wooded swamps that lie at the margins of the floodplain

Gt2—this is a veneer deposit of glacial till lying on a 120 m high outcrop of dolomite

Bedrock Units

Rq—these are 65, 100 and 80 m high outcrops of Lower Cambrian quartzite, the largest being mined for crushed stone

Rd—are outcrops of Lower Cambrian dolomite

Photo Source

Courtesy of National Air Photo Library, A30382, 174-175

References

Field Geology Services (2006). Fluvial geomorphology assessment of the lower Winooski River, Vermont.

Stewart, D. P (1974). Geology for environmental planning in the Milton-St Albans region, Vermont. Vermont Geological Survey.

Stewart, D. P., & MacClintock, P. (1969). The surficial geology and Pleistocene history of Vermont. Vermont. *Geological Survey, Bulletin 31*.

Figure S3: Gandoman

Location 31°51′N, 51°06′E

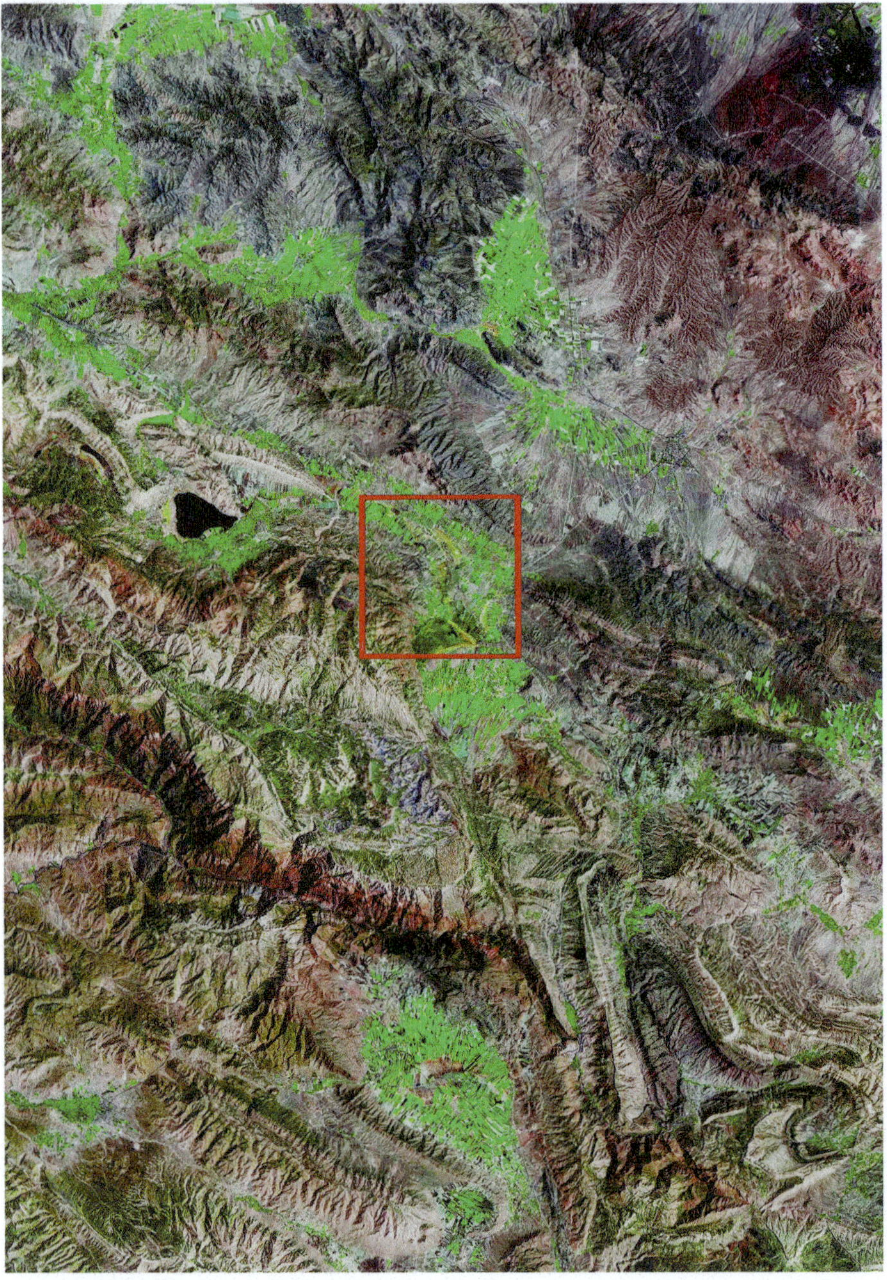

Landsat Image T/M 4-5, 04 August 2009

Regional Environment

The scene shows the Example area to be one of a number of spectral green intermont basins in western Iran. The area is in an arid zone with 500 mm annual precipitation.

Structure

The site area is in a belt of imbricate folds of Mesozoic and Cenozoic sedimentary rocks near the Main Zagros Thrust in the Zagros Mountains Tertiary Orogen of western Iran.

Physiography

The site is an intermont basin or *playa* with a 3 km^2 body of standing water at 2,215 m elevation. Surrounding highland rises to 2,600 m. Sediments of three Quaternary epochs occur in the basin area. It is filled with thick deposits of fluvio-lacustrine sediments and lies at the south end of a 50 km long strike valley segmented by alluvial fan dams. Occurring morphologic units are differentiated by topo site and land use.

Fv
L
FU2
FU2
FU2
W
L
W
W
ful
L
L
R2
W
L
ful
R2
ful
W
L
L
W
S
d
fv
S
W
S
W
R2
ful
L
S
L
R2
L
ful
fv
L
R1
W
s
L
FU2
ful
W
W
R1
FU2

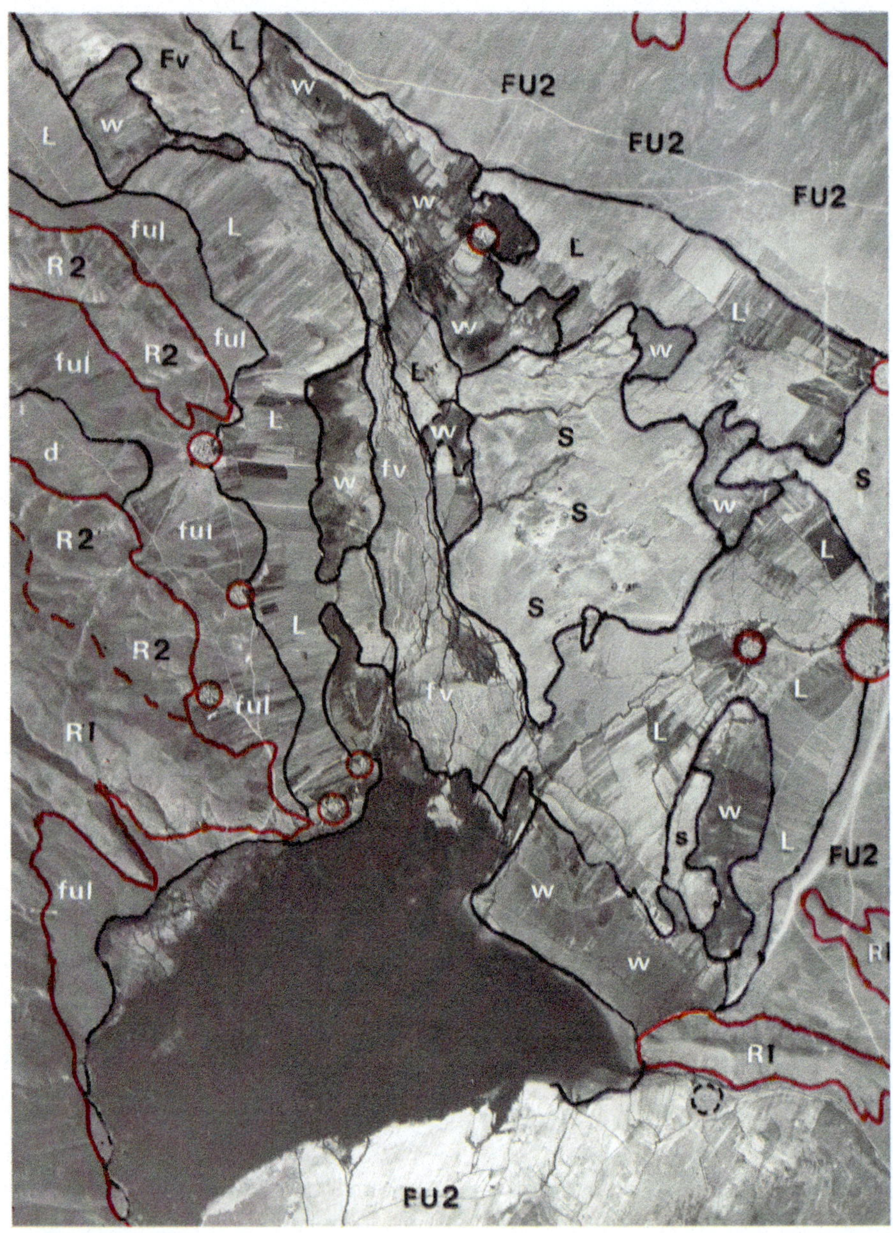

S3 Airphoto

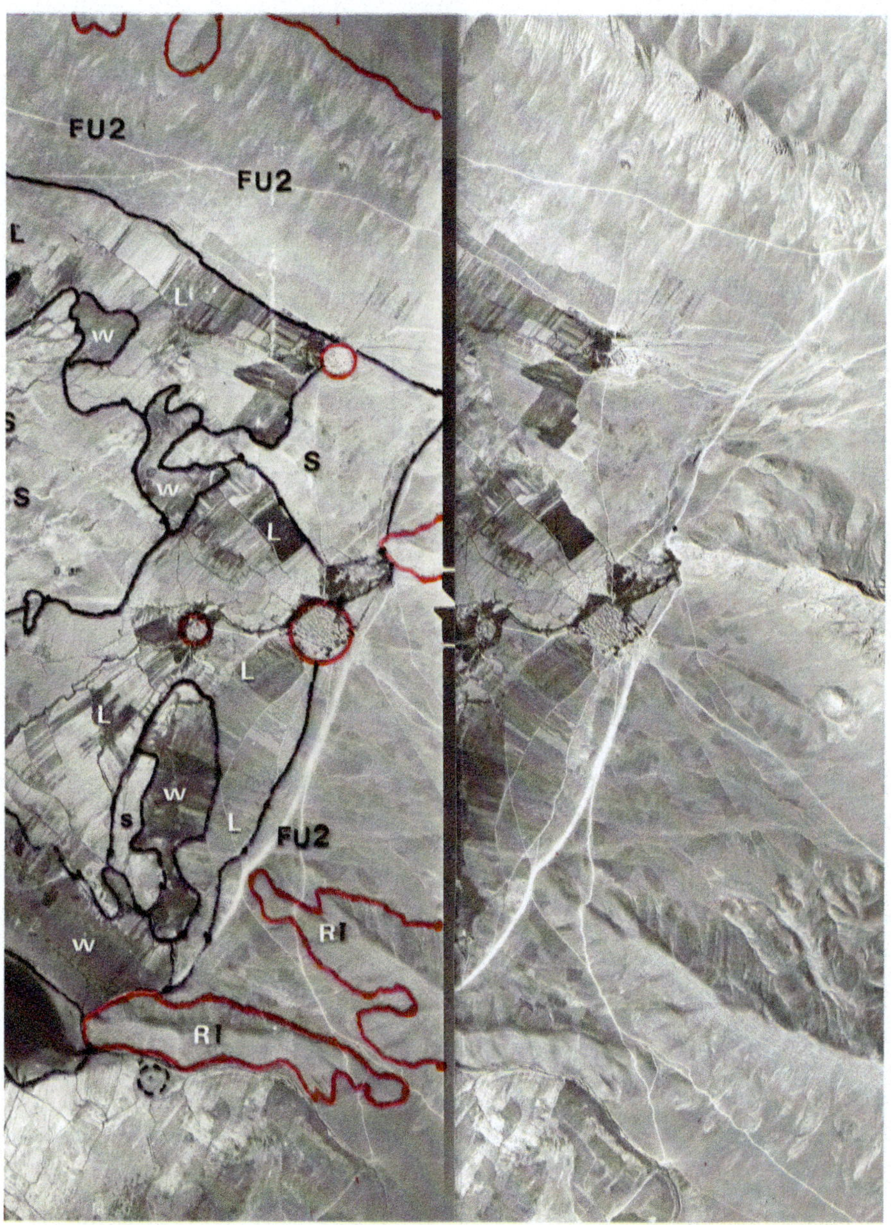

S3 Stereopair

Photo Interpretation

The 1:50,000 scale photo covers 88 km^2. Seven units of surficial deposits and two units of bedrock occur in the photo area. The air photo dates from circa 1955; Google Earth™ imagery of 2011 shows the mud of the lake bottom with no standing water.

L—These are silt and clay *Pleistocene* pluvial lacustrine deposits. They rise in 5 km from lake level to 15 m terraces, and are completely occupied with irrigated agriculture.

S—these are uncultivated salt pan remnants of the *Holocene* playa

W—These are high groundwater wet zones now fully irrigated. The largest area near the lake has been brought into cultivation.

Fv—These are *Recent* alluvial deposits of a perennial stream that flows from a catchment area in the mountains to the northwest of the region. The stream cuts through the lacustrine beds as it flows toward the lake. Irrigation canals maintain floodplain crops.

Fu1—These are alluvial fans sloping down from 2,260 m elevation to the lacustrine flats on west side of the basin margin. The fans are also under irrigated agriculture.

Fu2—this is an uncultivated alluvial piedmont apron (*bajada fan*) sloping down to the northeast side of the basin from 2,500 m

D—a small debris flow at the base of the marlstone hills

Bedrock Units

R1—these are Mesozoic calcareous mountains that regionally enclose the basin and are at 2,400 m in the photo area

R2—these are low weathering hills of marlstone associated with the higher mountains

Note:

Red circles indicate eight villages that surround the basin. The larger circle on the right is the town of Gandoman.

Photo Source

Personal archive

References

Currey, D. R., & Sack, D. (2009). Hemiarid Lake Basins: Hydrographic patterns (pp 471–487). In A. J. Parsons, & A. D. Abrahams (Eds) *Geomorphology of desert environments*. Heidelberg: Springer.

Mabbutt, J. A. (1969). Desert Lake Basins. In *Desert landforms* (pp 180–214). Cambridge: MIT Press.

Shaw, P. A., & Thomas, D. S. G. (1989). Playas, pans and salt lakes, pp 184–205. In D. S. G., Thomas (Ed) *Arid zone geomorphology*. London: Belhaven Press, Halsted Press.

Figure S4: Webb

Location 50°14′N, 108°16′W

Landsat Image T/M 4-5, 26 October 2011

Regional Environment

The area is in the Interior Plains of southwest Saskatchewan.

Physiography

The site is at an altitude of 800 m elevation on the dry, grass-covered glaciated southern interior plains. Sites **A** and **B** are sand dune complexes located downwind from sand in post glacial lakes in southeast Alberta. Site **A** is a northeast oriented ridge lying between a glaciolacustrine area on the south and a glaciofluvial area on the north.

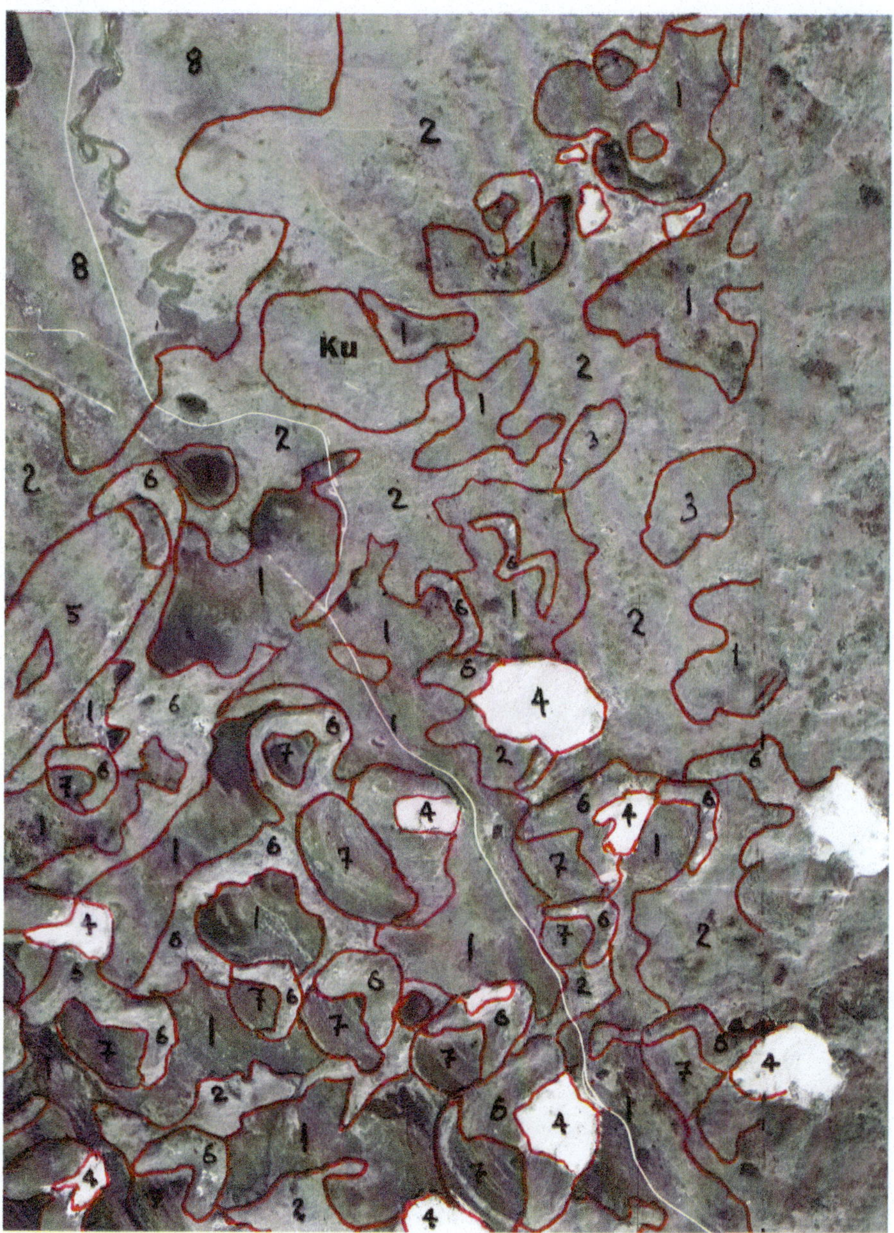

S4 Airphoto

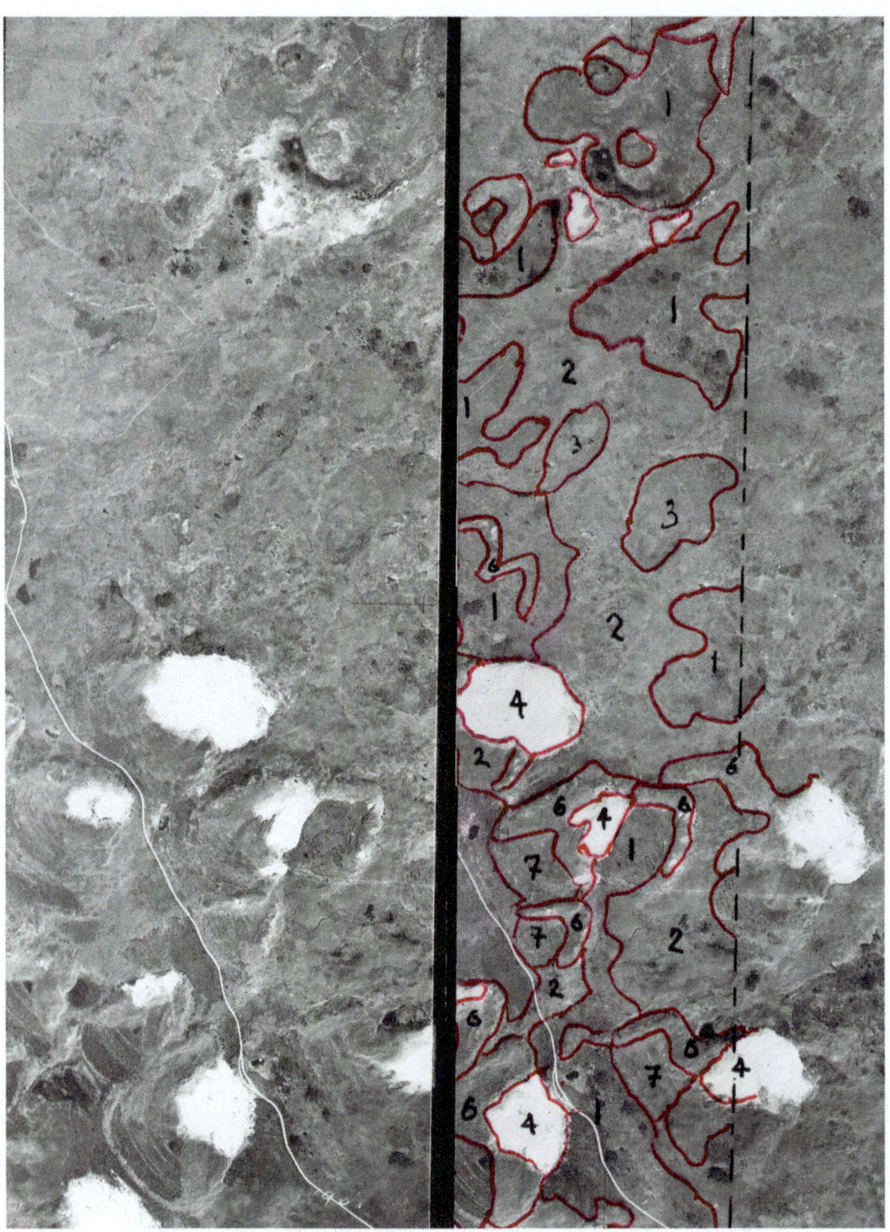

S4 Stereopair

Photo Interpretation

The 1:20,000 scale air photo area covers 14 km^2. Six Aeolian, two postglacial and one bedrock deposit occur in the site.

Dunes

These consist of medium to fine sand and coarse silt that is well sorted.

5—linear this is a single kilometer long ridge 3 m high on its right side, striking northeast on the west margin of the photo

6—parabolic These are numerous in the south half of the site. They are generally barer of grass cover. They were formed by winds from the southwest.

7—beach lines these are strand lines of small ponds that were on the concave side of the parabolic dunes

4—blowouts these bare units are depressions that result from wind erosion on the parabolic dunes

Inter Dune Units

2—Sand sheets these are extensive low hummocky deposits of sand without dune forms in the north and east of the site

3—deflation hollows these are old depressions, partly back-filled, in the sand sheet surfaces

Non Aeolian Deposits

1—these are flat, shallow glaciolacustrine deposits amid the dune system

8—this is part of the glaciofluvial deposit that borders the dune complex on the north

Ku—is a low sand-covered outcrop of Upper Cretaceous shale and minor sandstone

Photo Source

Courtesy of National Air Photo Library, A14967, 16-17

References

Maathuis, H., & Simpson, M. (2007). *Groundwater resources of the Prelate (72 K) area, Saskatchewan*. SRC Publication No. 11975–1EO7.

Landslide Terrains

Twenty-one landslide Examples are classified into seven types of movements:
- rock avalanches, **E1**
- rock slides, **E2**
- rock slumps, **E2** through **E11**
- debris avalanches, **E12** through **E17**
- debris slides, **E18** and **E19**
- retrogressive earth flows, **E 20** and **E21**

Occurrences are in seven terrain environments:
- polar—three Examples
- glaciated alpine—eleven Examples
- glaciomarine—two Examples
- glaciated plains—one Example
- volcanic tephra—three Examples
- crystalline plateau—one Example
- graben faulting—two Examples

Nearly half the Examples are in glaciated alpine terrain. The prevalence of landslides in such environments, particularly debris slides and debris avalanches is due to the concurrence of physical weathering, steep slopes and high relief, J. Gerrard (1990) states "*Debris flows and slides are so conspicuous on steep mountain slopes that it is likely that they are the main mechanism for long-term slope evolution.*" (Mountain Environments, MIT Press 1990, p 86).

- Case histories of 137 landslides in European alps are given in G. H. Eisbacher and J. J. Clague, (1984). Destructive mass movements in high mountains: Hazard and management, Geological Survey of Canada Paper 84–16.
- A recommended report is Highland, L. M. and Bobrowski Peter, *The Landslide Handbook – A Guide to understanding landslides,* United States Geological Survey Circular 1325, 2008, 129p.

Figure E1: Ekalugad

Location 68°52′N, 69°22′W

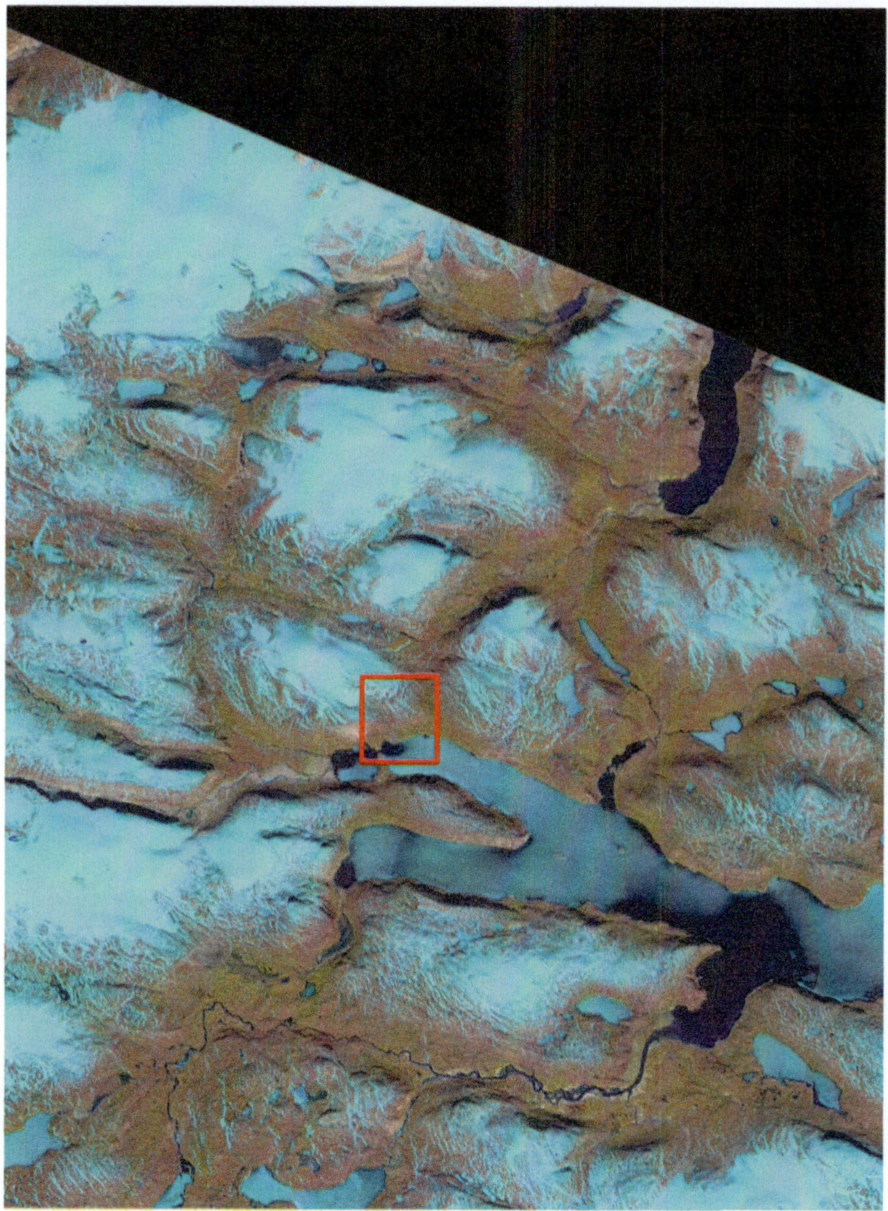

Landsat Image T/M 4-5, 05 July 1996

Regional Environment

The image covers 1,800 km^2 of the east coast of Baffin Island of the northeast Canadian Shield. Most of the terrain was still snow covered at date of image acquisition.

Structure

The site is at the head of the 55 km long Ekalugad Fjord that is incised into the Eocene continental rifted margin of the Northeast Baffin Shelf of Baffin Bay. Structurally controlled valleys divide the area into blocks of granite and gneiss with small ice caps occurring on summits above 1,000 m.

Physiography

The valley west of the site and that of Tingin Fjord to the northeast contain braided floodplains and deltas of glaciofluvial outwash gravels and sands.

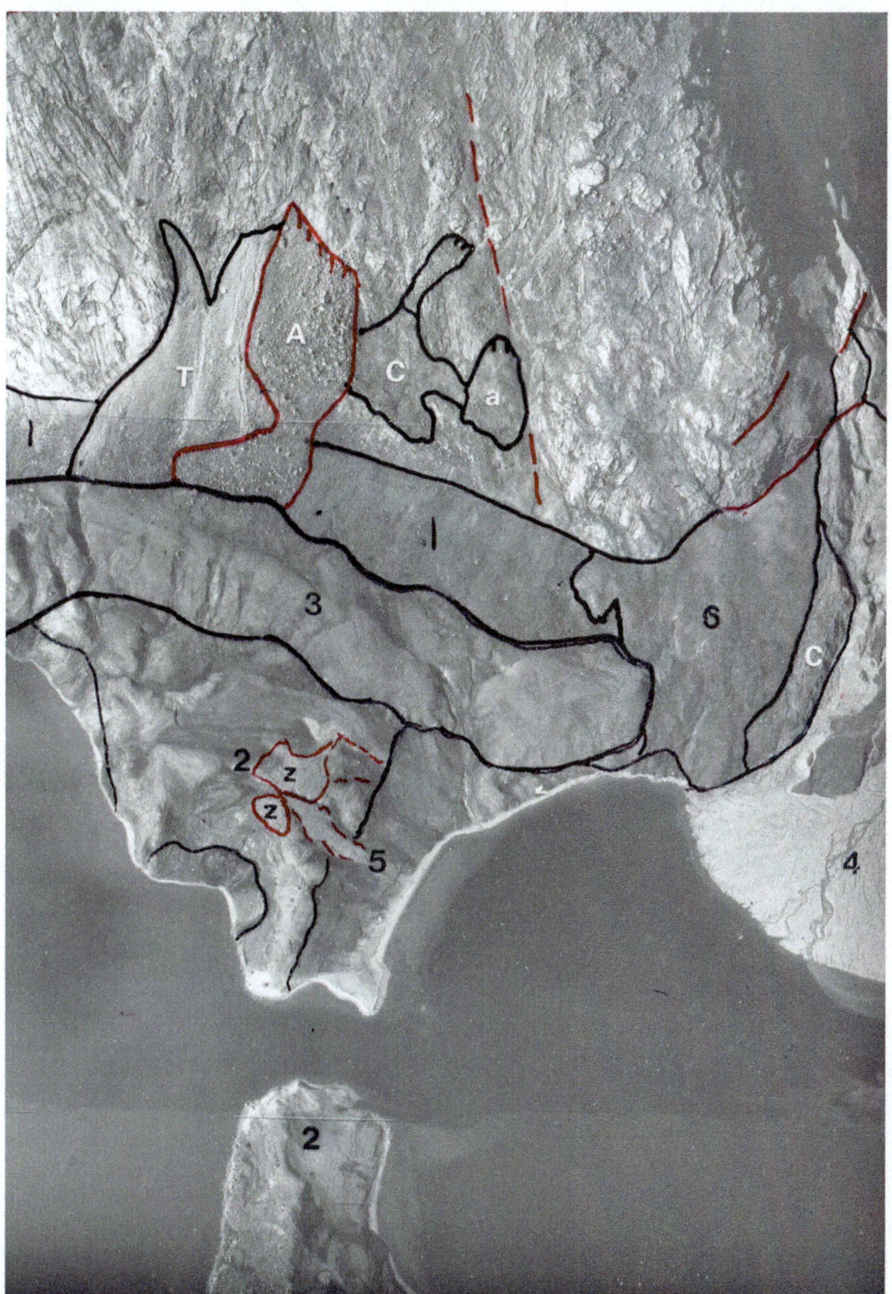

E1 Airphoto

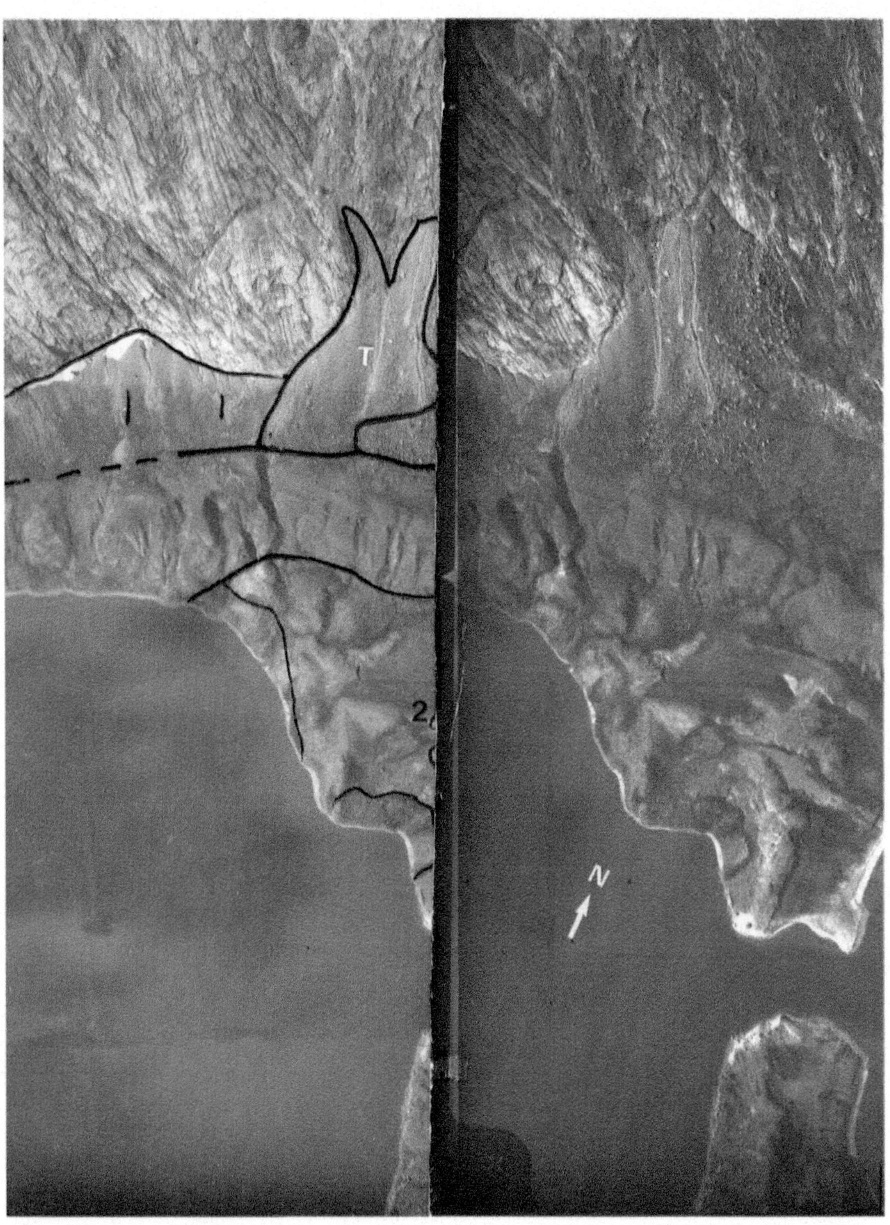

E1 Stereopair

Photo Interpretation

Eleven geounits of three categories occur in the 1:25,000 scale photo area.

Landslides

A—Is a kilometer long 400 m wide rock block slide of angular boulders and cobbles that slid along a 200 m sheet surface from a crown which is part of a red traced set that mark the edges of local granite sheets which controlled the block mass detachment. The boulders are strewn along the slope in a coherent repose state with a few carried onto the marginal moraine. The crown of the unit is at 685 m elevation and the toe is at 175 m.

T—is a 400–900 m wide by 1,200 m long debris flow with characteristic levees along abandoned gullies.

C and **a** are debris slides

Glacial Units

1—is glacial till of a marginal moraine

2—is a terminal moraine

3—is a kame terrace of stratified sand and gravel

5—are glaciomarine sediments.

Fluvial Unit

4—is an active fan-delta.

Periglacial Units

6—are gelifluction sheets on marginal moraine

z—are ice wedge polygons

Comment

The general area in which this example is located is discussed as the Home Bay Upland on pages 167–171 of *A Report of Physiographic conditions of Central Baffin Island and adjacent areas, RAND Corporation Memorandum RM -2837-PR, January 1963*. The account in the report does not provide detailed site information to be referenced.

Photo Source

Courtesy of National Air Photo Library, A20199, 49-50

Figure E2: Sloko

Location 59°05′N, 133°40′W

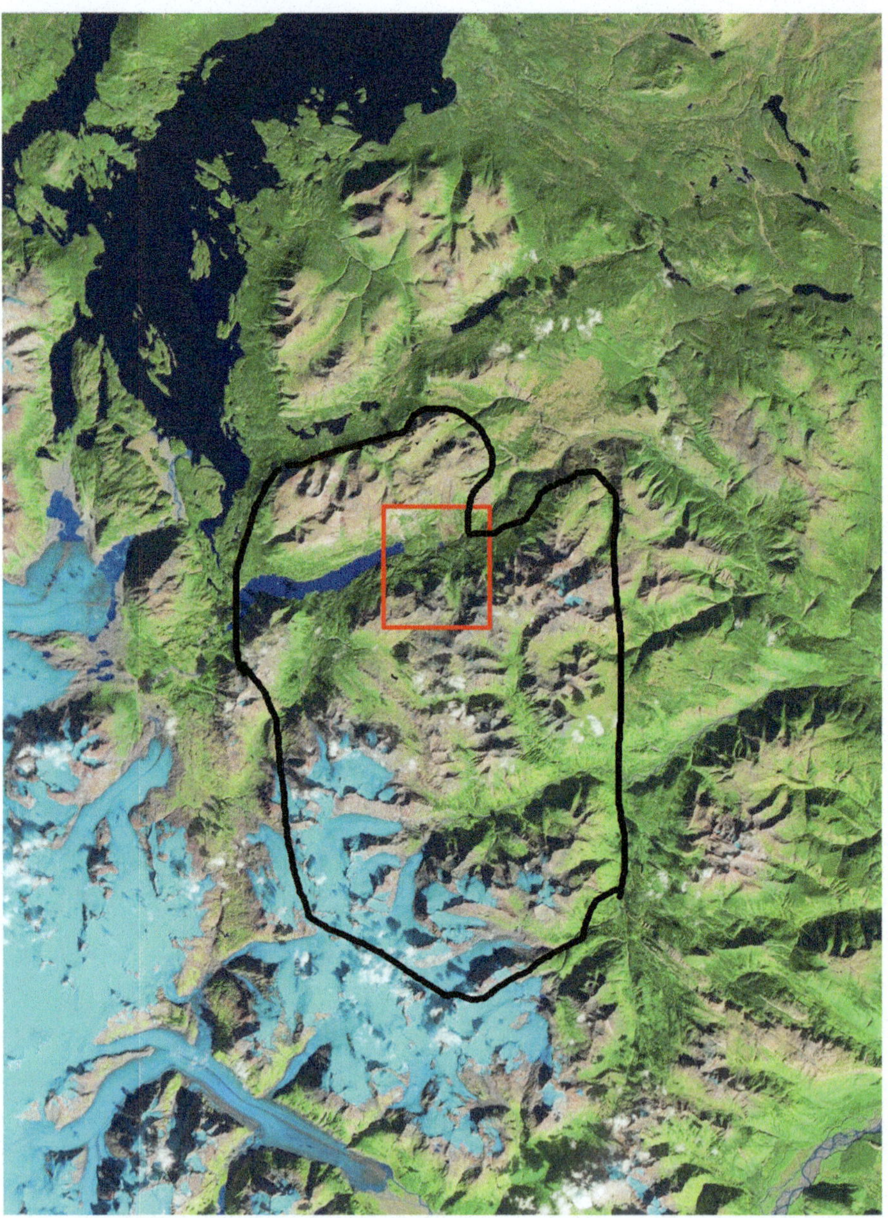

Landsat Image T/M 4-5, 17 September 1995

Regional Environment

The scene covers 3,000 km^2 of the Tagish Highlands of northwestern British Columbia. The site lies between the south end of Atlin Lake in the north and the blue 2,400 m high ice caps of the Boundary Ranges in the south.

Structure

The site is part of the Sloko Lake Volcanic Complex (SLVC), outlined in black, which is the largest erosional remnant (600 km^2 in area) of a Paleogene volcanic complex within the northern Canadian Cordillera. It is located along a linear trend of volcanic plutonic rocks that once formed part of a widespread continental volcanic arc that extended from west-central British Columbia into southeastern Alaska and southwestern Yukon.

Physiography

The volcanic complex area is glacially dissected. Elevations in the north range from 700 m in valleys to 1,500 m peaks. The glacier ice field area in the south is part of the Coast Mountains with elevations of 1,600–2,000 m.

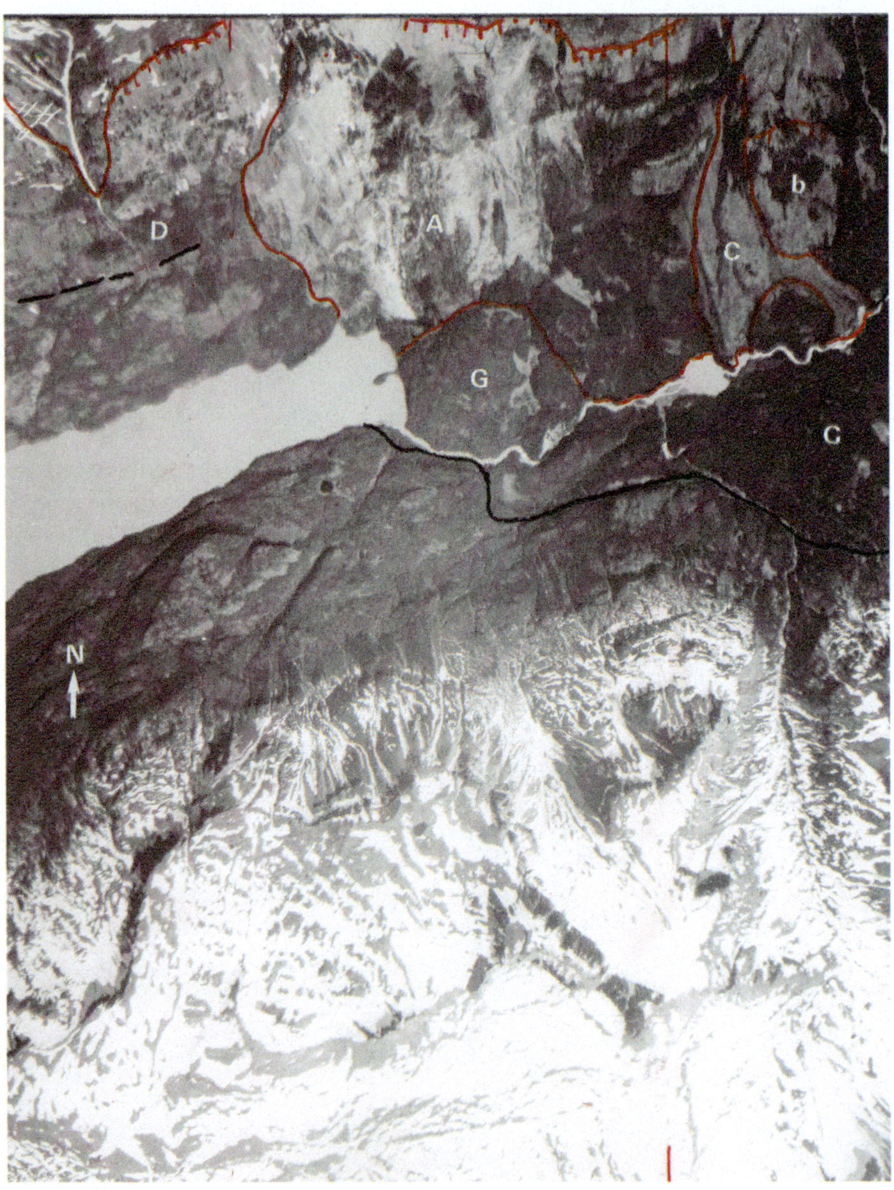

E2 Airphoto

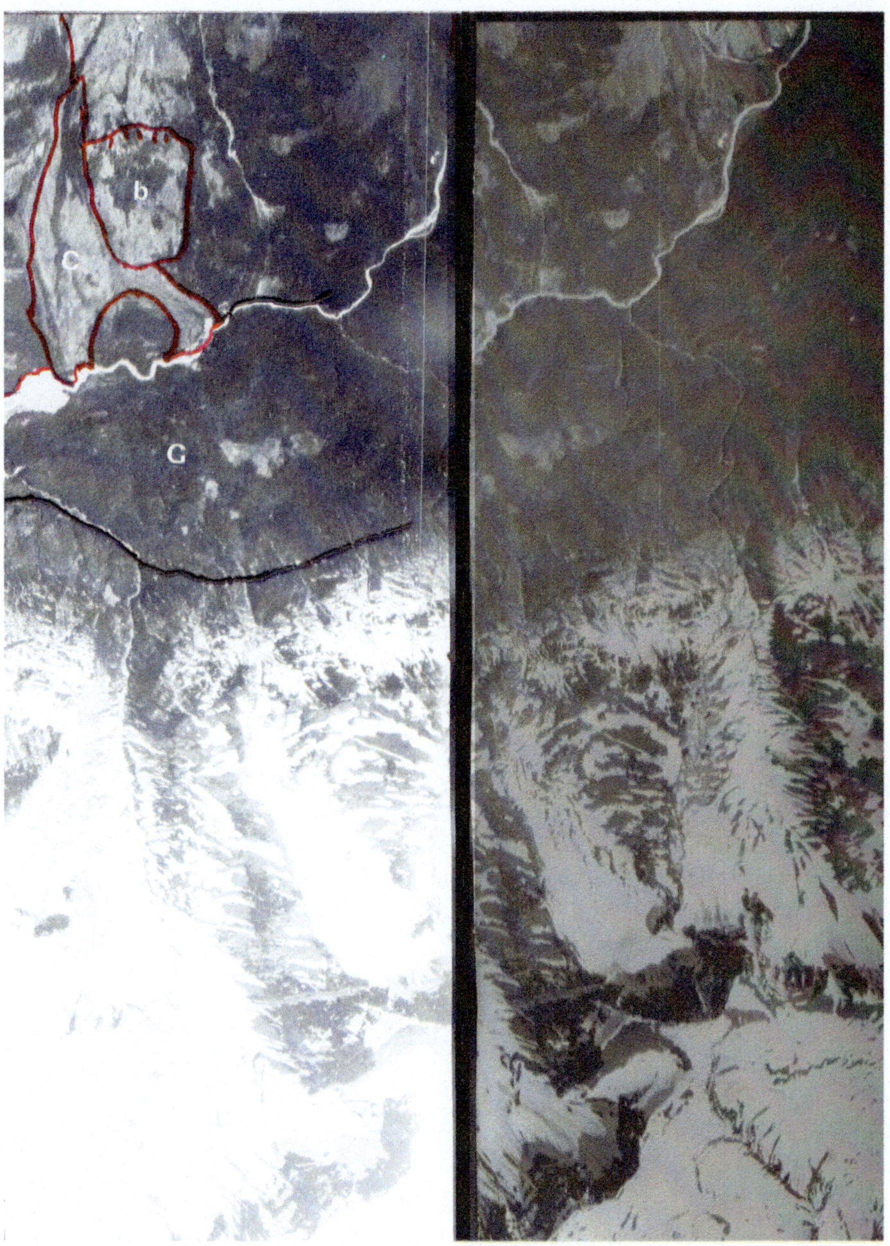

E2 Stereopair

Photo Interpretation

Four types of landslides appear adjacent to one another in this 1:40,000 scale air photo of the 1 km long east end of 725 m elevation, glacier meltwater fed 12 km long Sloko Lake.

The **A** rock slump is a 2,100 m wide by 850 m long series of blocks of displaced material on a steep slope lying on the surface of rupture below the bare main scarp. The active part is distinct from the wooded older adjacent masses. The step-like blocks may be rotational and probably relate to the interbedding of weak tephra and resistant lavas.

b—is a partly wooded rock slide

C—is a debris slide with characteristic levees along abandoned stream courses

D—is part of a complex of slumps and slides that occur for a distance of 5 km westward along the same slope that occurs on the lineament-related north shore of the lake. The same volcanic rocks occur on both sides of the lake but the unstable slope area relates only to the lineated north shore.

G—indicates glacial valley morainic deposits which dam the lake

Comment

All the slides appear to be in states of recurrent activity.

Photo Source

Courtesy of National Air Photo Library, A11385, 25-26

Reference

Petrogenesis of the Paleogene Sloko Lake Volcanic Complex, Northwestern British Columbia and applications for early Tertiary magmatism in the Coast Plutonic Complex. MSc thesis Resnick J. (2003) Department of Earth and Planetary Sciences, Mc Gill University, Montreal, Quebec.

Figure E3: South Cape Fjord

Location 76°35′N, 84°55′W

Landsat Image T/M 4-5, 13 September 2006

Regional Environment

The image covers 1,200 km^2 of half ice cap-covered South Cape Fjord on southern Ellesmere Island.

Structure

The area belongs to the Arctic Platform plateaux of Lower Paleozoic unfolded carbonate and clastic rocks of the southeast Canadian Arctic Islands.

Physiography

Local ice caps are at about 1,000 m elevation. The 2 km wide 25 km long Sydkap Glacier of the 45 km long fjord has receded 8 km in 12 years from its position in the map of 1994. The fjord empties into Jones Sound in the southeast 25 km from the site.

This fjord coast line mimics major northwest striking fault systems of the Heim Peninsula to the south of the photo area.

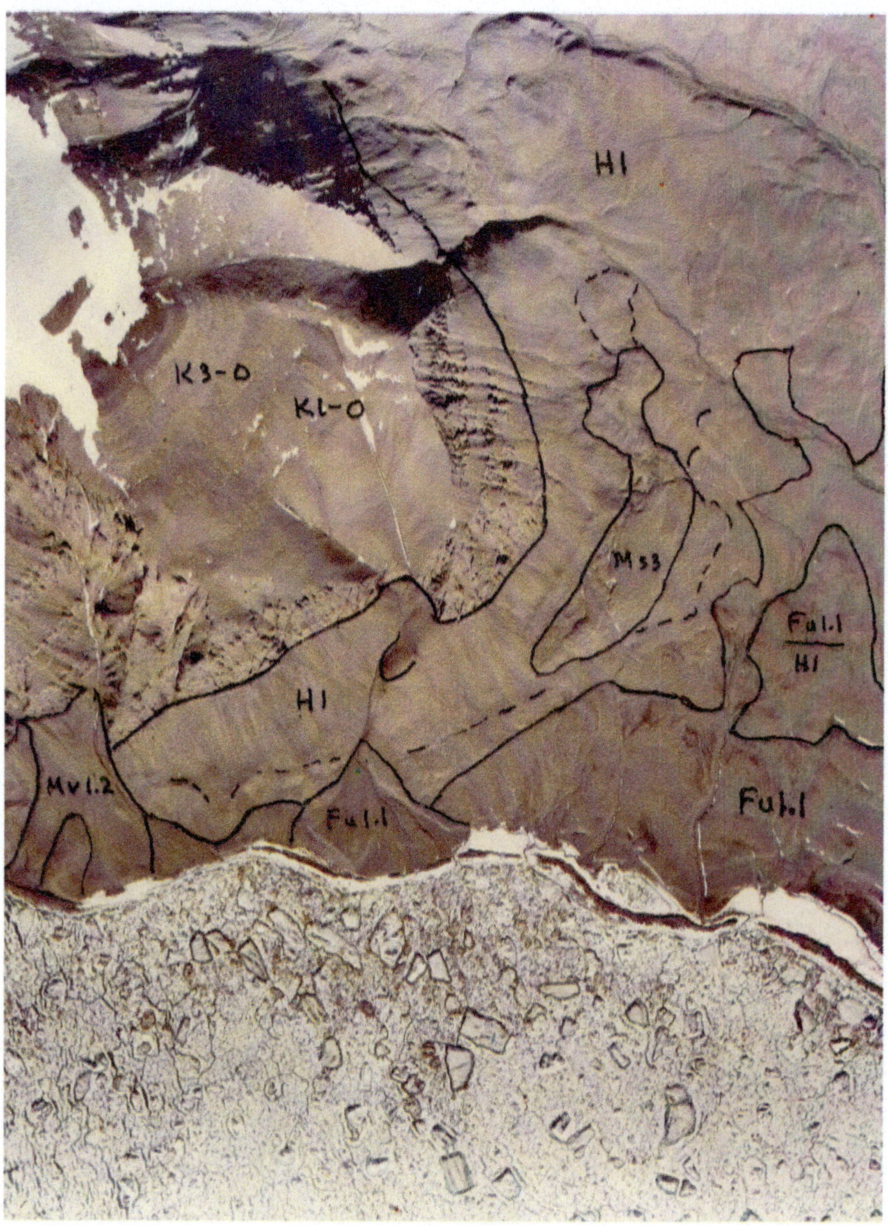

E3 Airphoto

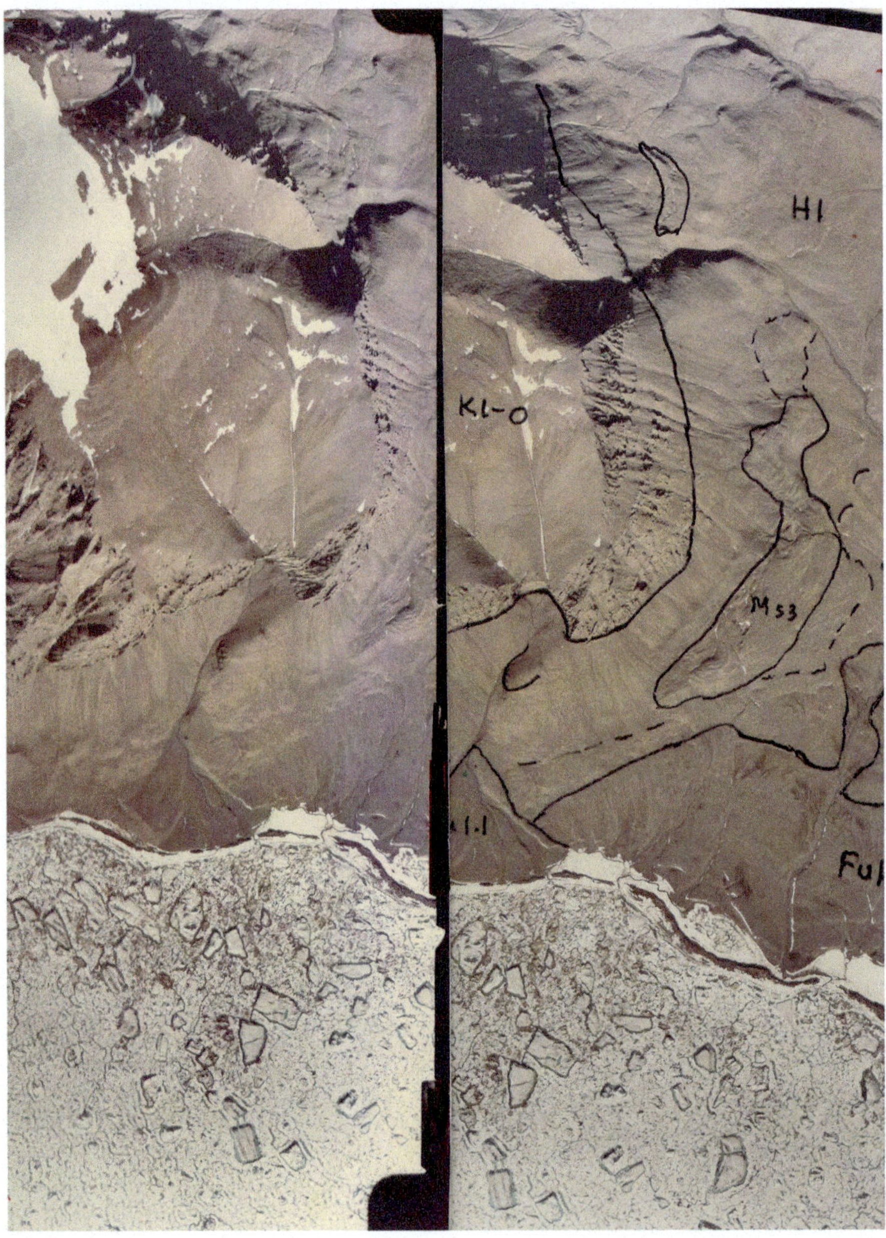

E3 Stereopair

Photo Interpretation

The site on this 1:20,000 scale natural colour photo consists of three mass movement types and three bedrock types.

Mass Movements

The **Ms3** slump mass is 1,200 m long by 250 m wide and is approximately 30 m in height. It appears to have moved 155 m down on a moderate slope without back tilting from the contact of the H1 evaporites below the limestone escarpment at 330 m elevation.

The **Fu1.1** unit is part of a 10 km broad fan-delta.

Mv1.2 are debris flows

Ordovician Bedrock Types

H1—is evaporate dolostone

K1-O—is limestone

K3-O—is dolostone and limestone

Photo Source

Courtesy of National Air Photo Library, A31000, 167-168

Reference

GSC Map1840A, *Baad Fiord-Cardigan Strait*, 1:250,000, 1994.

Figure E4: North Tanquary 1

Location 81°38′N, 76°38′W

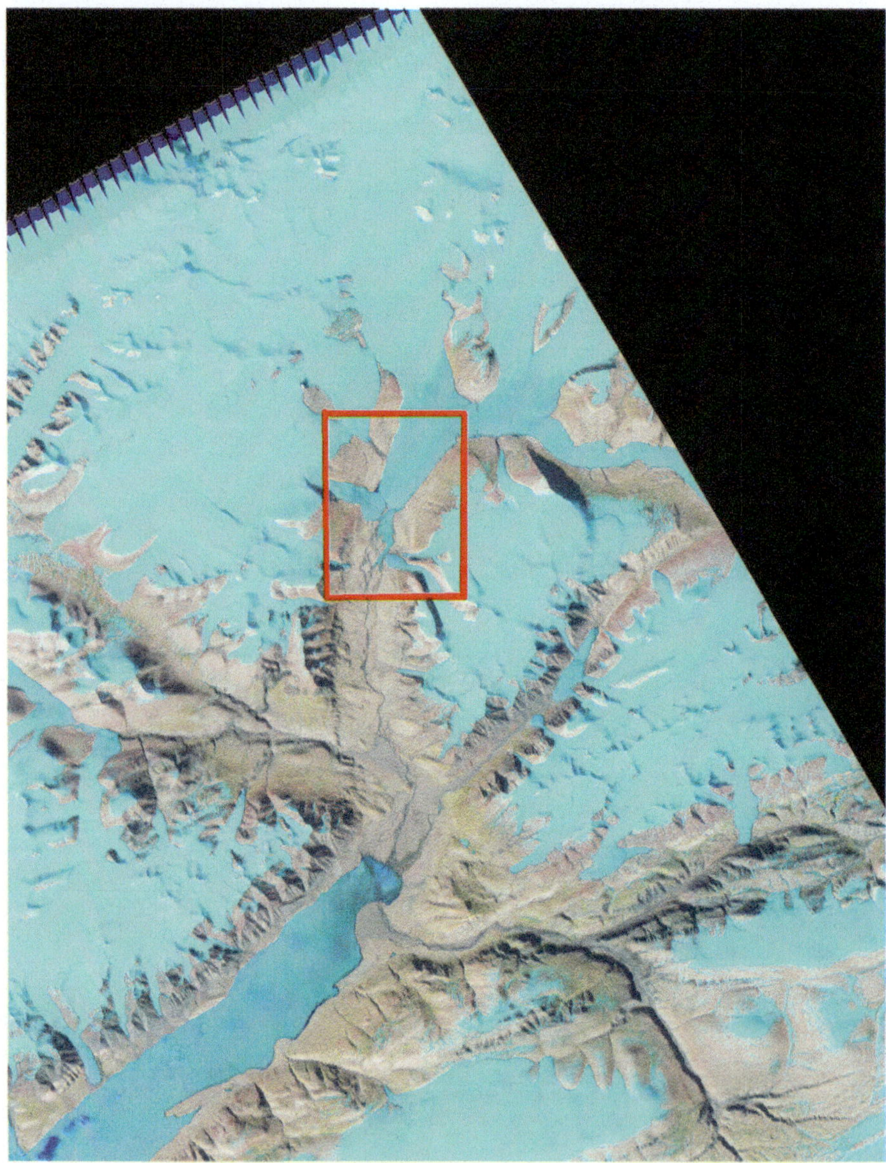

Landsat Image L7, S2C–on, 19 June 2002

Regional Environment

The image shows the Example site to be in one of a group of valleys tributary to the head of Tanquary Fjord among surrounding 1,400 m high ice caps of the Grant Land Mountains in northern Ellesmere Island of the Queen Elizabeth Islands, Canadian Arctic archipelago. The area is 150 km north of Figure B2.

The structure and physiography of this Example are further discussed in the description of Figure B2.

Structure

The Example is located near the north end of Late Paleozoic sediments of the mainly Cretaceous Svedrup Basin. The basin is flanked on both east and west by Mobile belts of Lower Paleozoic sediments.

Physiography

The site is a dissected upland of folded basin strata 20 km north of the north end of the fjord.

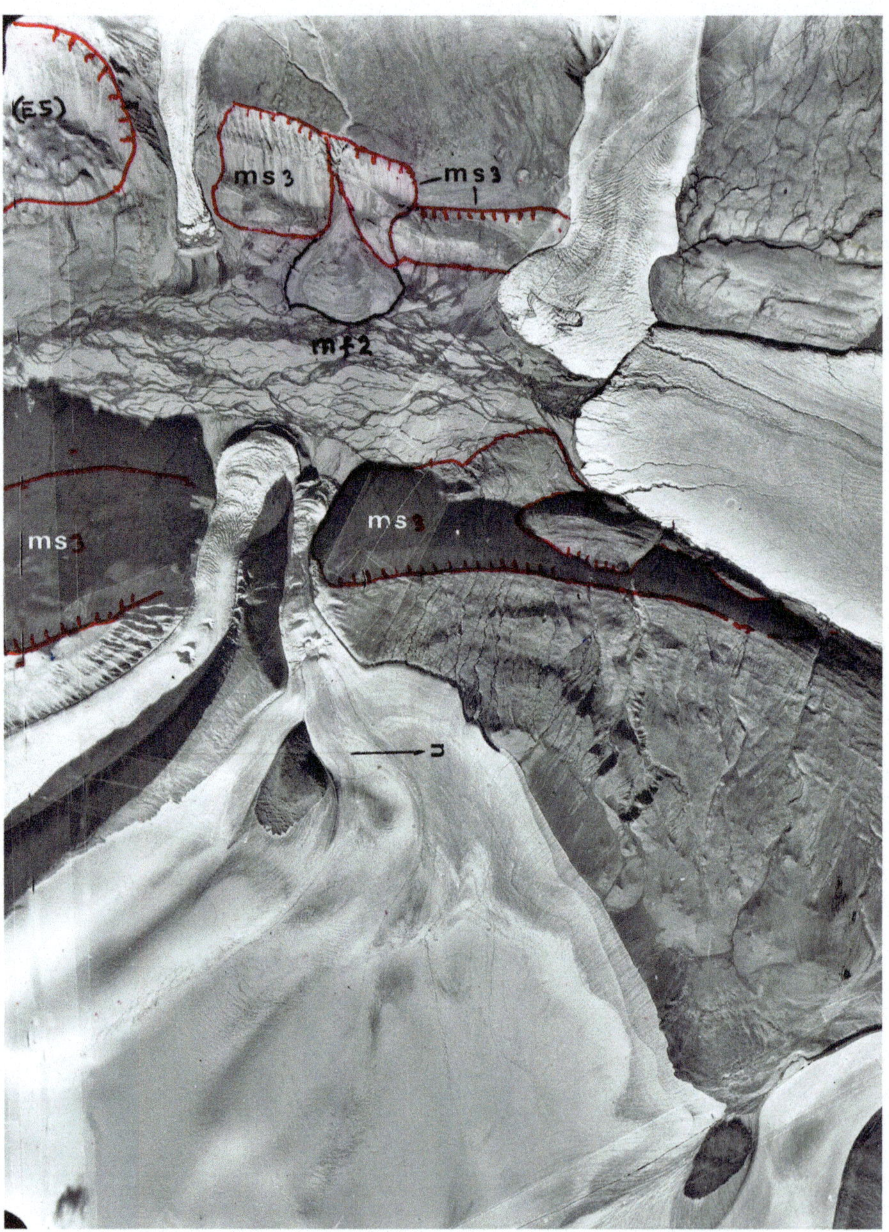

E4 Airphoto

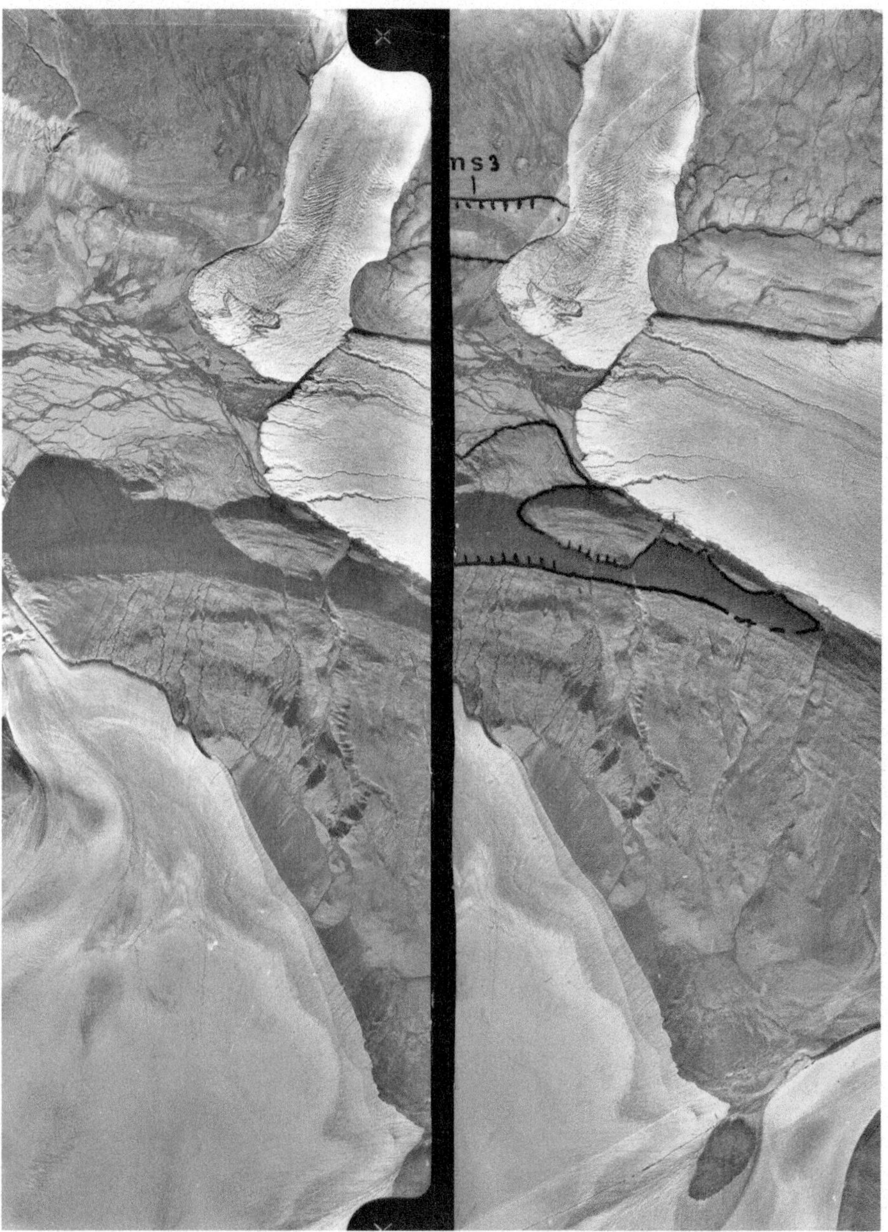

E4 Stereopair

Photo Interpretation

This site on this 1:60,000 air photo shows a number of **Ms3** rock slumps resting on both sides of the Air Force Glacier valley. They occurred in response to glacial retreat.

The slopes were oversteepened then debuttressed by the removal of lateral support by withdrawal of the glacier mass that occupied the valley The lower part and tongue of Air Force Glacier is now in the north half of the photo.

A single slump block showing little back-tilting rests at the base of the surface of rupture of each slide. The shadowed slumps also occur on the east side of the valley.

An **Mf2** earth flow with marked transverse ridges lies at the valley level of the braided valley stream.

Air Force Glacier is a subpolar glacier. Its terminus is in the same position as it was at date of photography 43 years ago.

The lower half of the smaller side glacier, touching Air Force glacier has extension crevasses of its descent of the steep slope from the ice cap. It has a bulging terminus of contorted ice which could indicate some surge activity. A small frontal moraine is visible.

Photo Source

Courtesy of National Air Photo Library, A16691, 56-57

Figure E5: Tanquary II

Location 81°38′N, 76°38′W

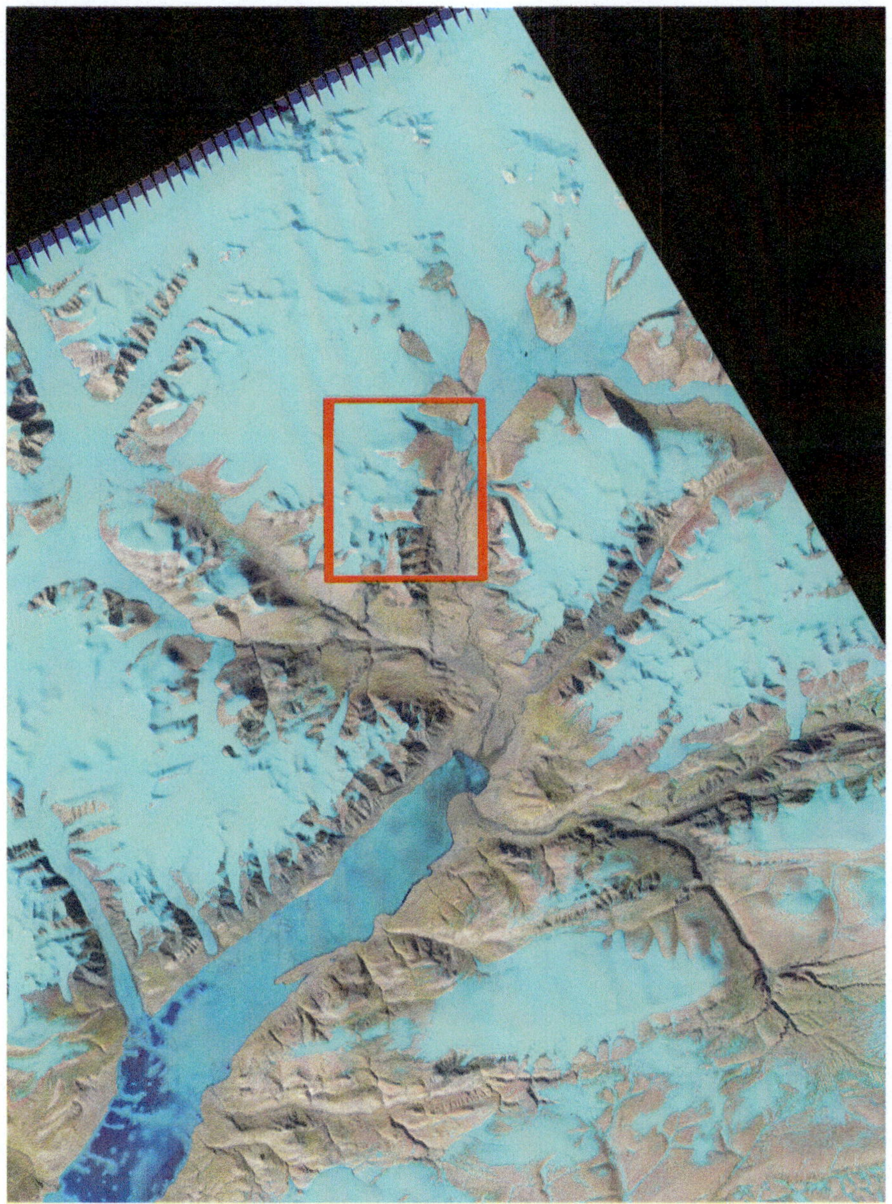

Landsat Image L7, S2C on, 19 June 2002

Regional Environment

This is the same image as that of Figure E4. Structure and physiography are an extension down valley of the terrain of E4.

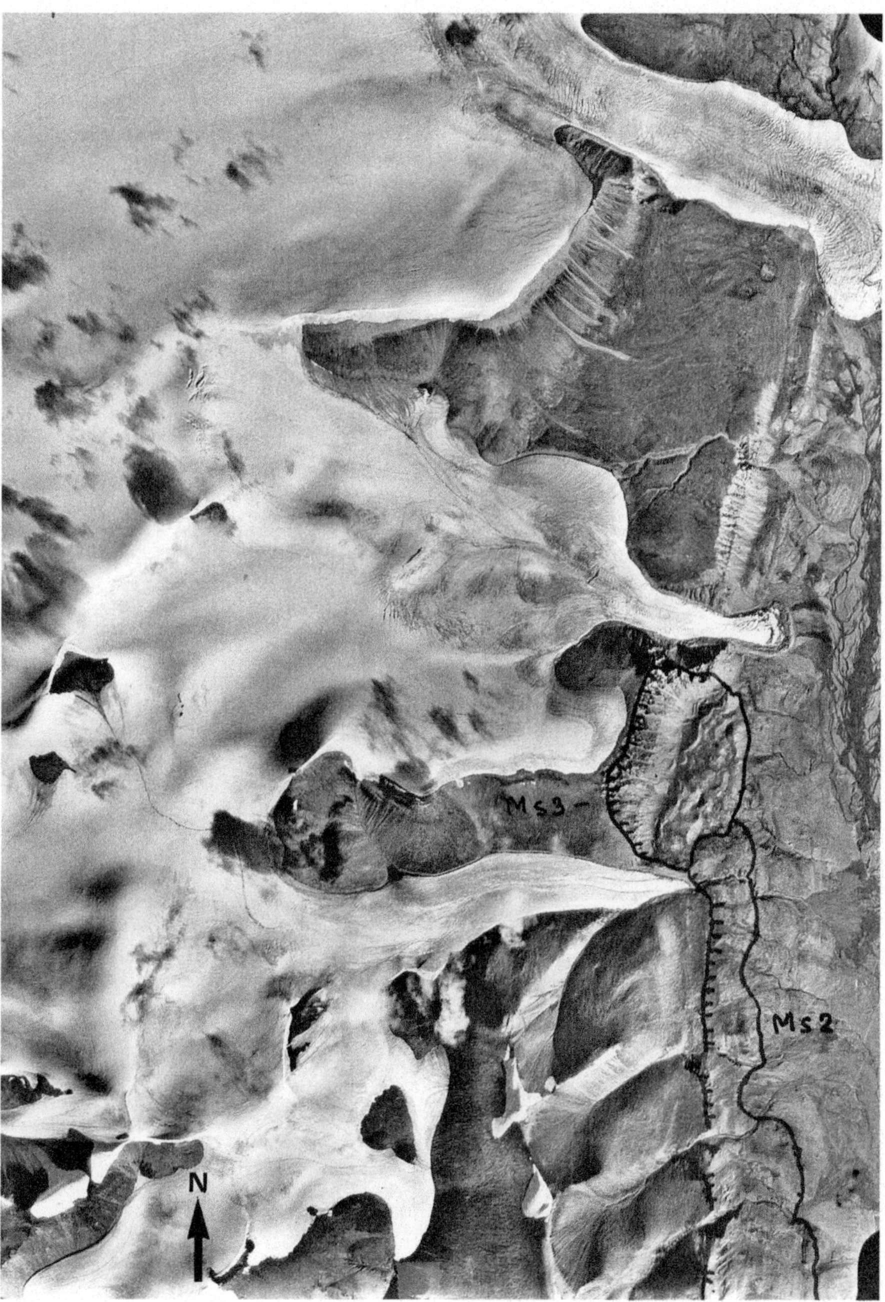

E5 Airphoto

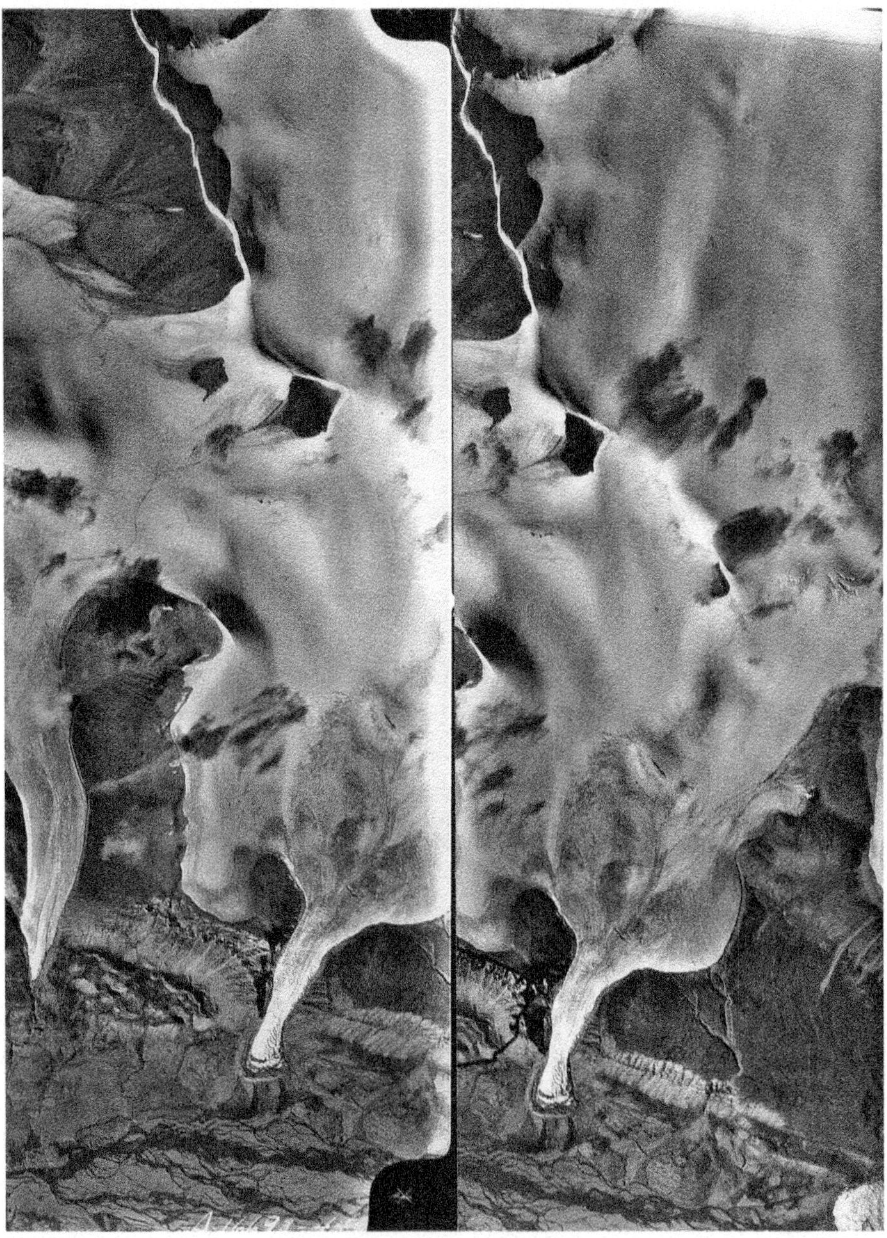

E5 Stereopair

Photo Interpretation

Air photo scale is 1:60,000

This **Ms3** slump, about 1 km south of those of E4, consists of a single disrupted mass of blocks with no apparent back-tilting resting at the base of the rupture surface.

A line of **Ms2** debris slides from a scarp lower than **Ms3** may be related to a particular shale bed.

Photo Source

Courtesy of National Air Photo Library, A16693, 44-45

Figure E6: Telegraph Creek

Location 57°51′N, 131°07′W

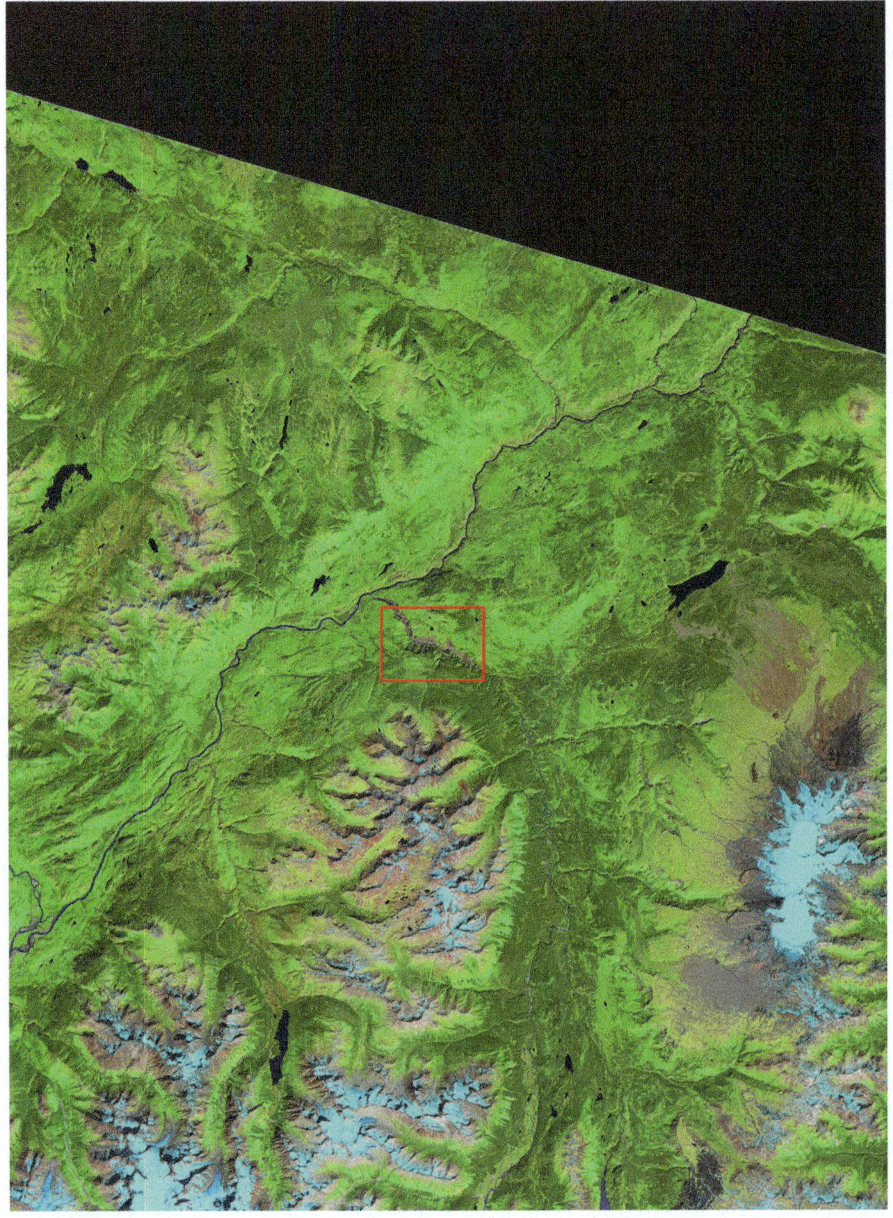

Landsat Image T/M 4-5, 05 August 2008

Regional Environment

The image covers 4,200 km^2. It shows the Example site to be in a valley tributary to the Stikine River which flows southwestward across the scene in the Interior Plateaux of northwest British Columbia. The site is 35 km northwest of Edziza Figure B9 on lower right.

Structure

The entire scene area is in the Intermontane geological belt of the Cordillera and is part of the Stikine accreted Terrane. The green areas are Upper Triassic volcanic and sedimentary rocks which are superposed by Cretaceous sedimentary rocks. The brown mass in the center is glacially dissected Carboniferous gneiss.

Physiography

The area is part of the Tahltan Highland unit of the Stikine Plateau. The relief within the Highland is controlled by the depth of incision of the Iskut River

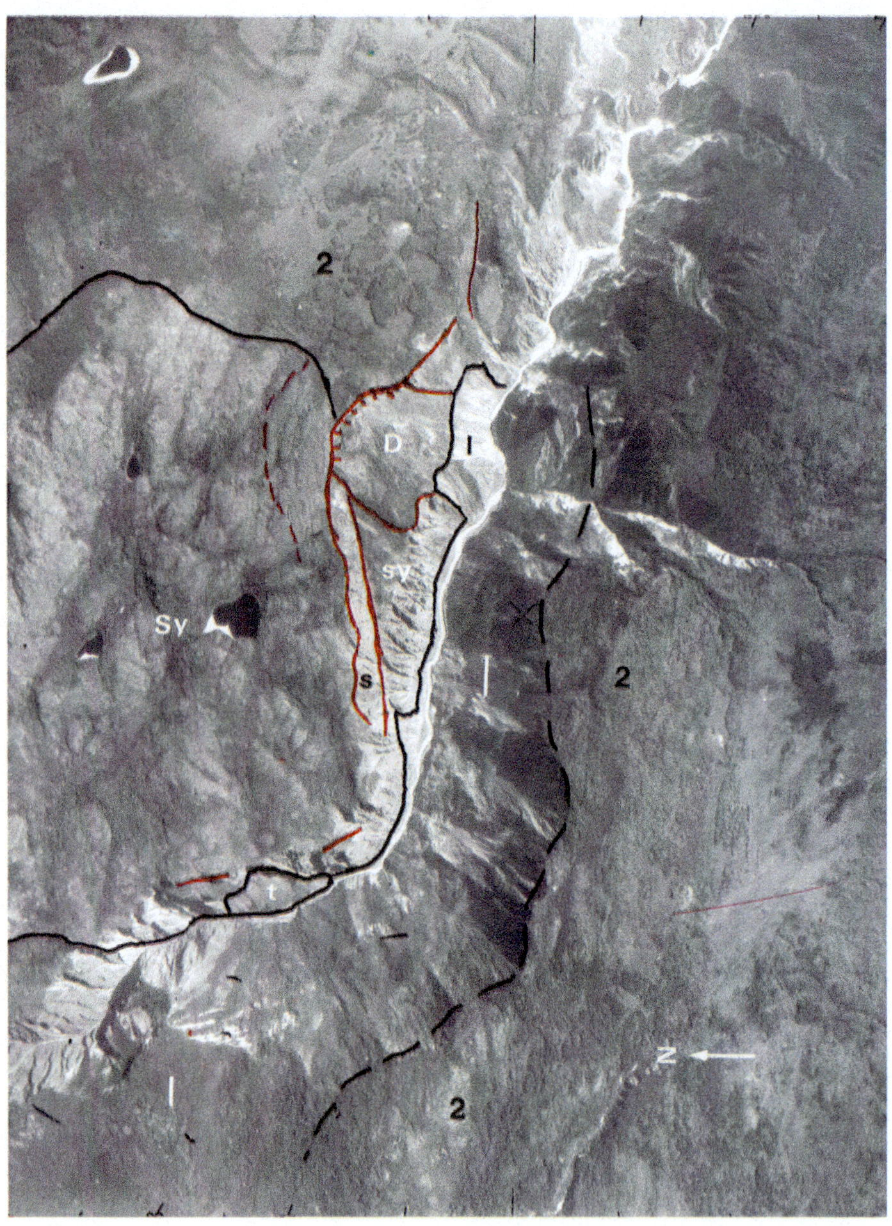

E6 Airphoto

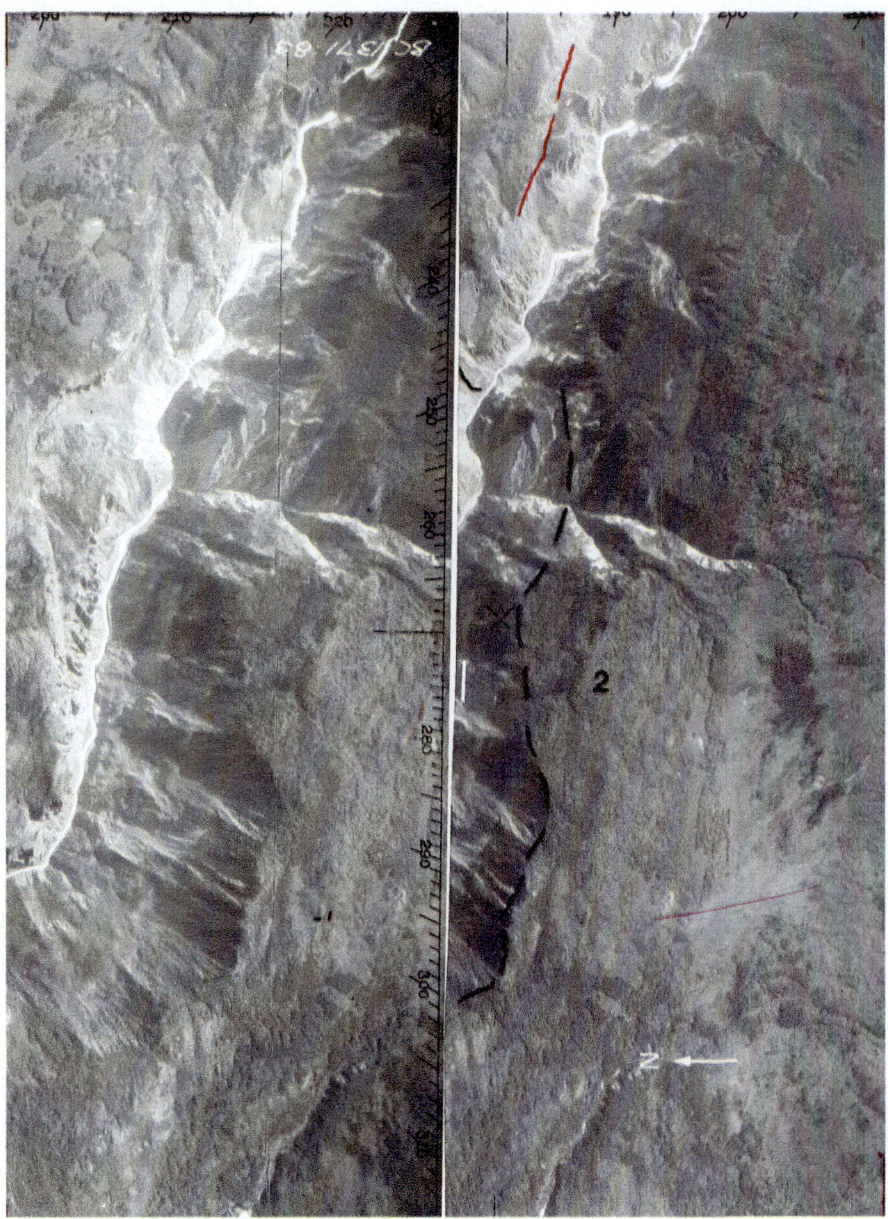

E6 Stereopair

Photo Interpretation

Three units of mass movement and three units of bedrock occur in the 1:40,000 scale air photo:

Mass Movements

s—is a fracture on the margin of the intrusion that has the appearance of a half-graben with a drop of about 130 m
D—is a meter wide debris slide from a basalt scarp
t—is an 800 m wide small talus deposit

Bedrock Units

1—are Upper Cretaceous sandstones and siltstones that infilled a fracture zone in Telegraph Creek
2—are Upper Triassic extensive basalts and associated clastic sediments
Sy—is a 4 km diameter 300 m high Triassic syenite stock

Photo Source

Personal archive

Figure E7: Morne Boeuf

Location 19°08′N, 72°14′W

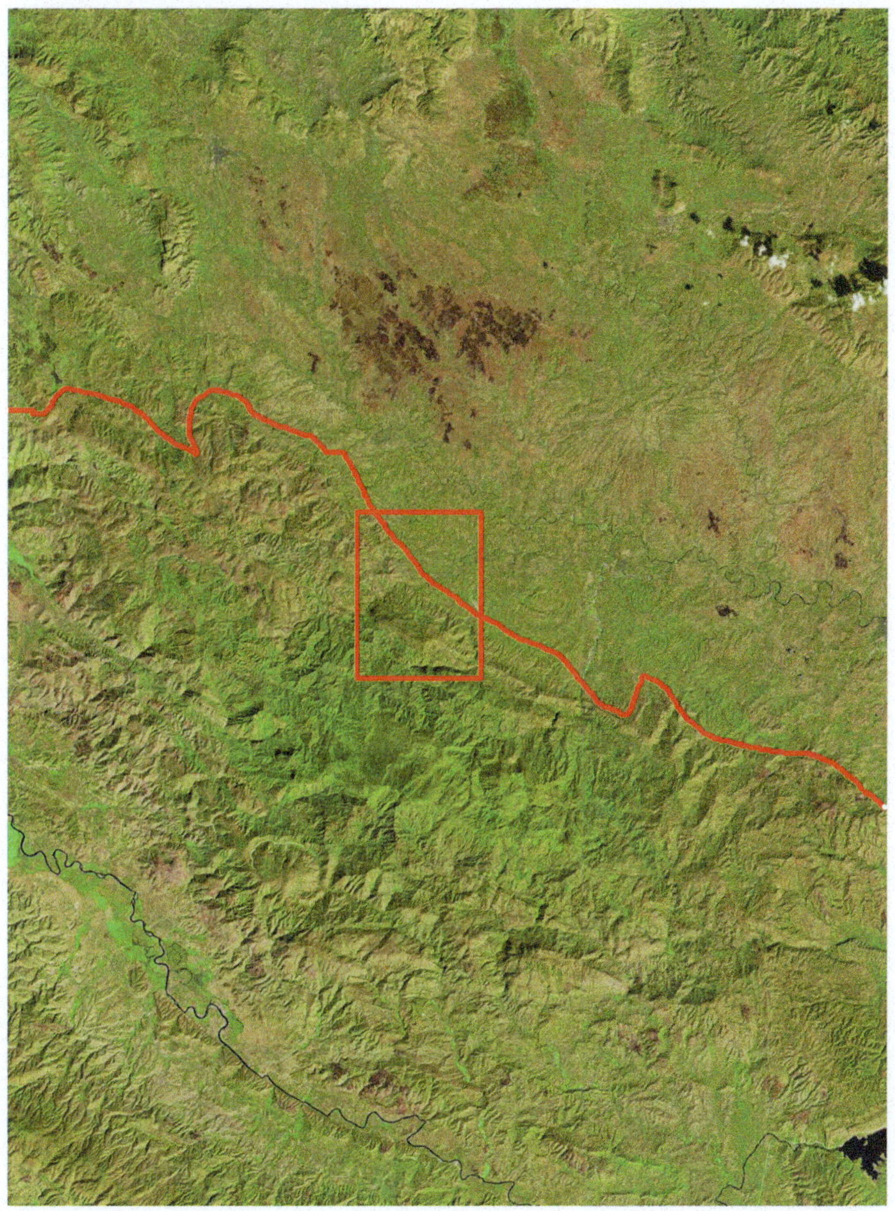

Landsat Image L7 S2C-on, 31 January 2002

Regional Environment

The image is in central Haiti in the Greater Antilles Deformed Belt of the Caribbean Plate.

Structure

The area is clearly divided by a red line into a northern plain and a southern mountain range. The Plain is composed of Miocene sandstones and marls; the dissected Montagnes Noires is an anticlinorium of Paleocene sedimentary and volcanic rocks.

Physiography

The north plain is at an average elevation of 350 m. The southern half is part of the 1,400 m Thomonde Chain of the Montagnes Noires north of the Artibonite valley. The anomalous dark brown zone in the plain are scub-covered zones of infertile Quaternary carbonate sediments.

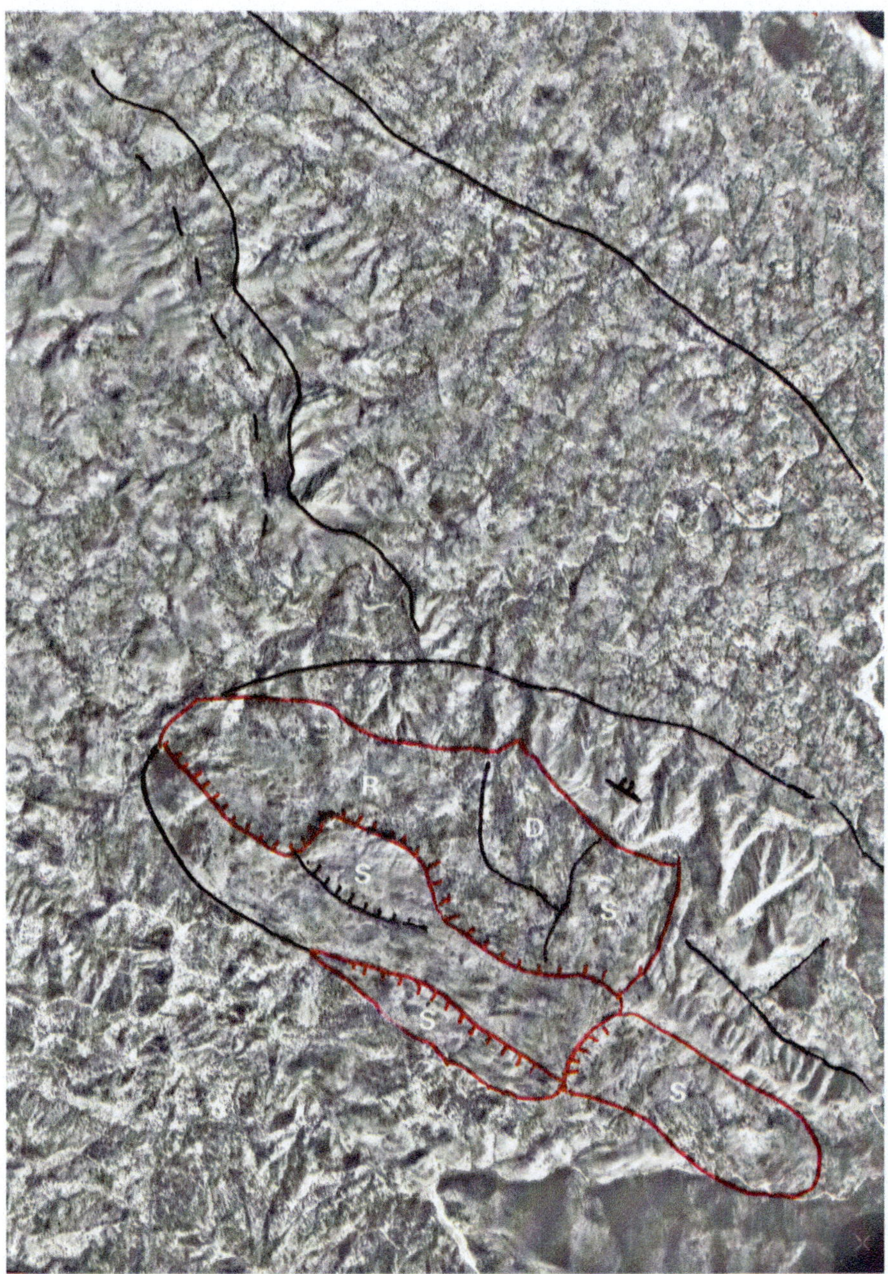

E7 Airphoto

E7 Stereopair

Photo Interpretation

The landslides are at 1,034 m elevation on a north slope of the west end of a 12 km long asymmetric antiform structure with steeper northern flanks. The structure lies on the north edge of the anticlinorium. The slides of Paleocene limestone are underlain by weak shales and siltstones.

R—this is the major failure; a 3 km long and one half to 1 km wide **rock slide**

S—this is a group of four **rock slumps** occurring on either side of the structure's crest

D—is a 700 m wide **debris slide** lying between the major slide and a 1,900 m long rock slump

Photo Source

Copyright IGN, 1978, HAI-400, 1113-1114

Reference

Munoz, N. G., Segovia A. V., Foss, J. E., (1980). Geology, morphotectonic analysis, and soils mapping of central Haiti, based on Landsat image and aerial photographs. In *Proceedings, 14th ERIM International Symposium on Remote Sensing*. pp 1529–1535.

Figure E8: Chiriqui

Location 08°44′ 46″N, 82°51′27″W

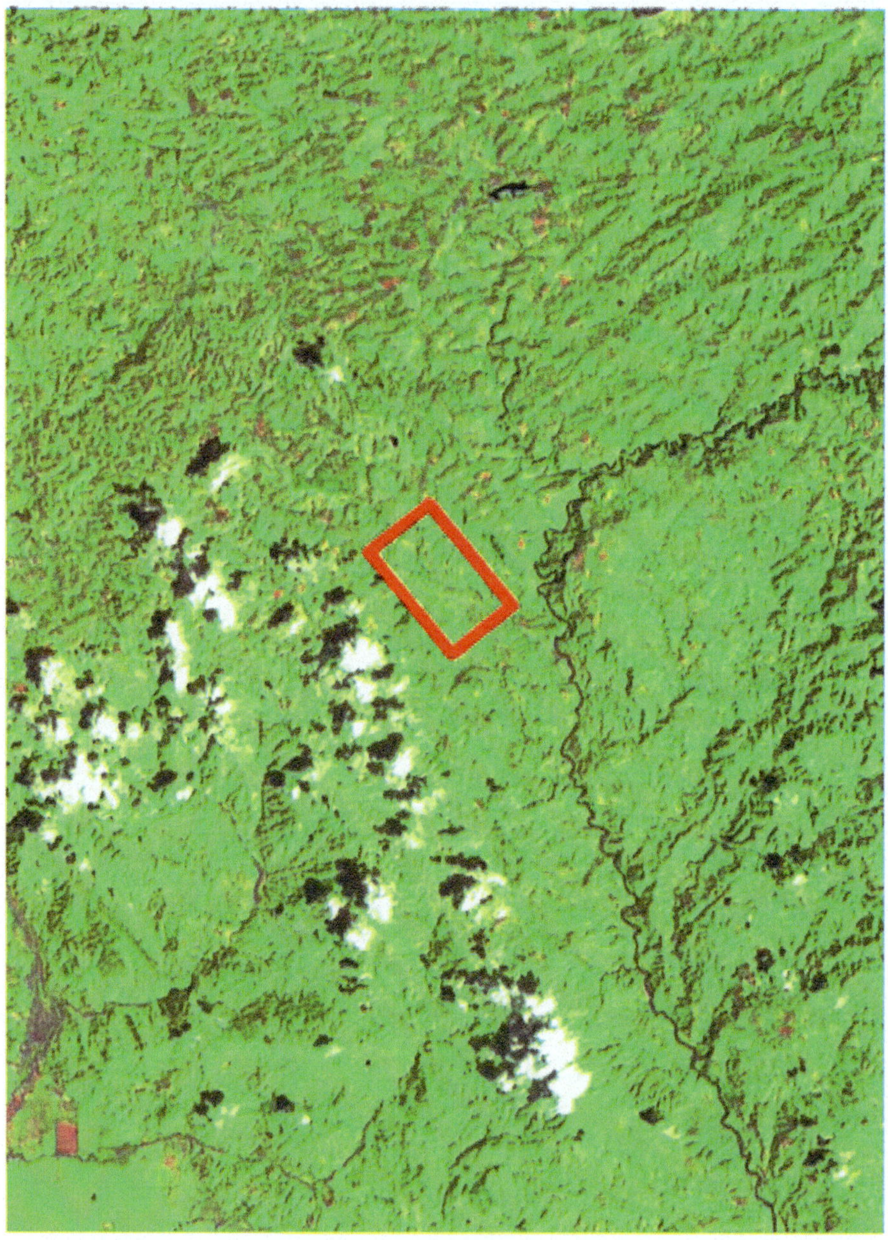

Landsat Image L7, S2C- on, 14 February 2000

Regional Environment

The image area is in western Panama just adjacent to the border of Costa Rica and is part of the Neogene Volcanic Belt of Central America along the boundary between the Cocos and Caribbean Plates. The ribbed appearance of the north part of the scene is dissected footslopes of tephra flows from a volcano to the northeast. The rest of the scene consists of flow sheets of tephra.

There is a record of regional seismic activity of 0–70 km focal depth.

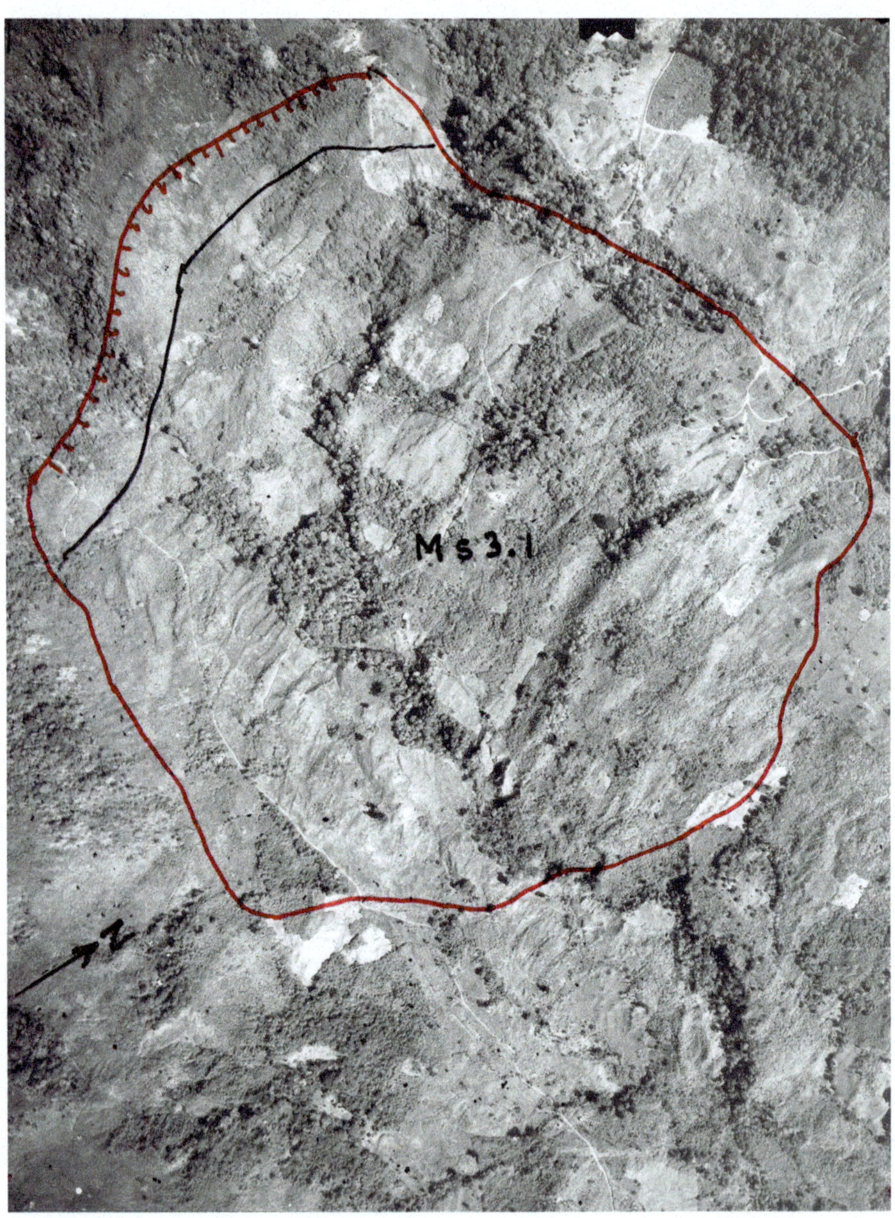

E8 Airphoto

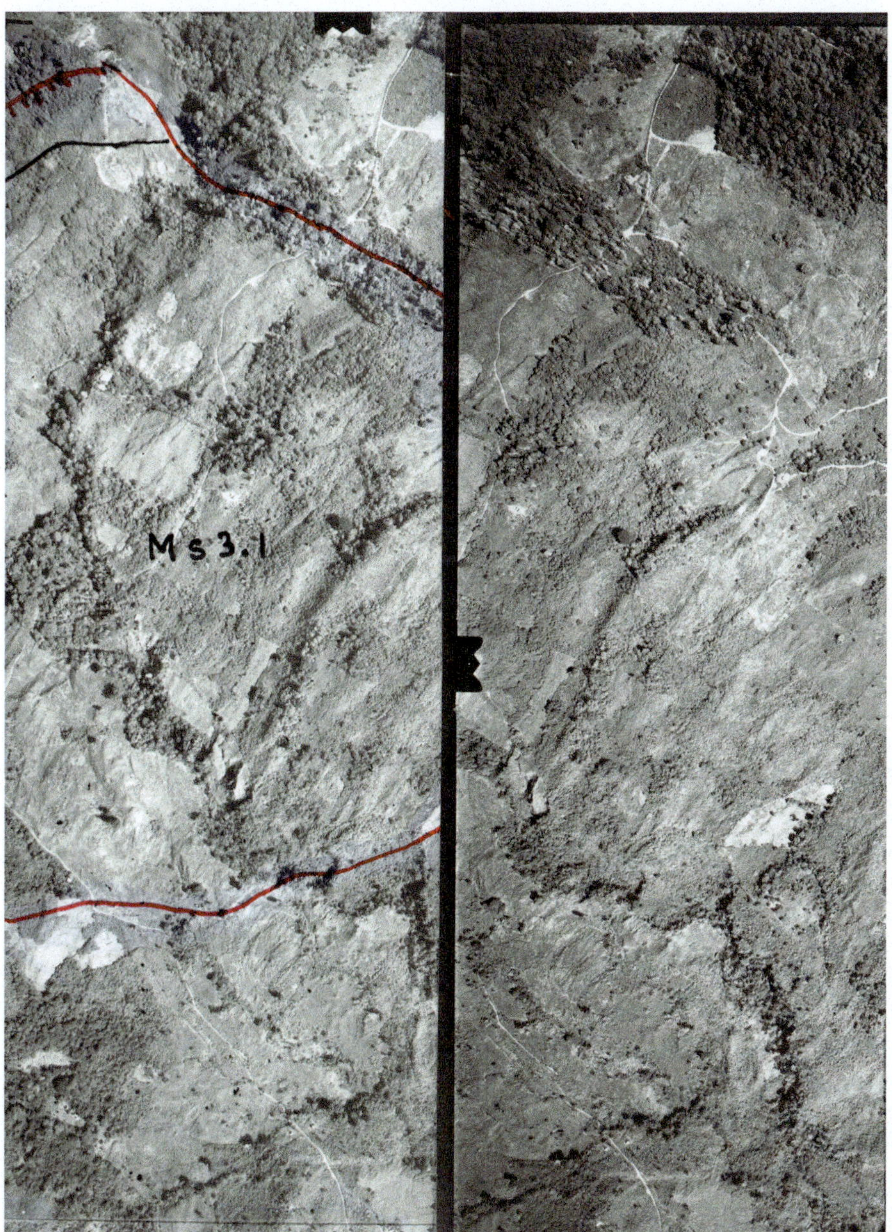

E8 Stereopair

Photo Interpretation

The classic characteristics of a slump slide are well expressed in this 1:10,000 scale air photo. The scarp of the slump is 10 m high below the crown with a 1,300 m long mass of backward tilting blocks of displaced material on the rupture surface.

Photo Source

Personal archive

Figure E9: Tabasara

Location 08°28′06″N, 81°35′54″W

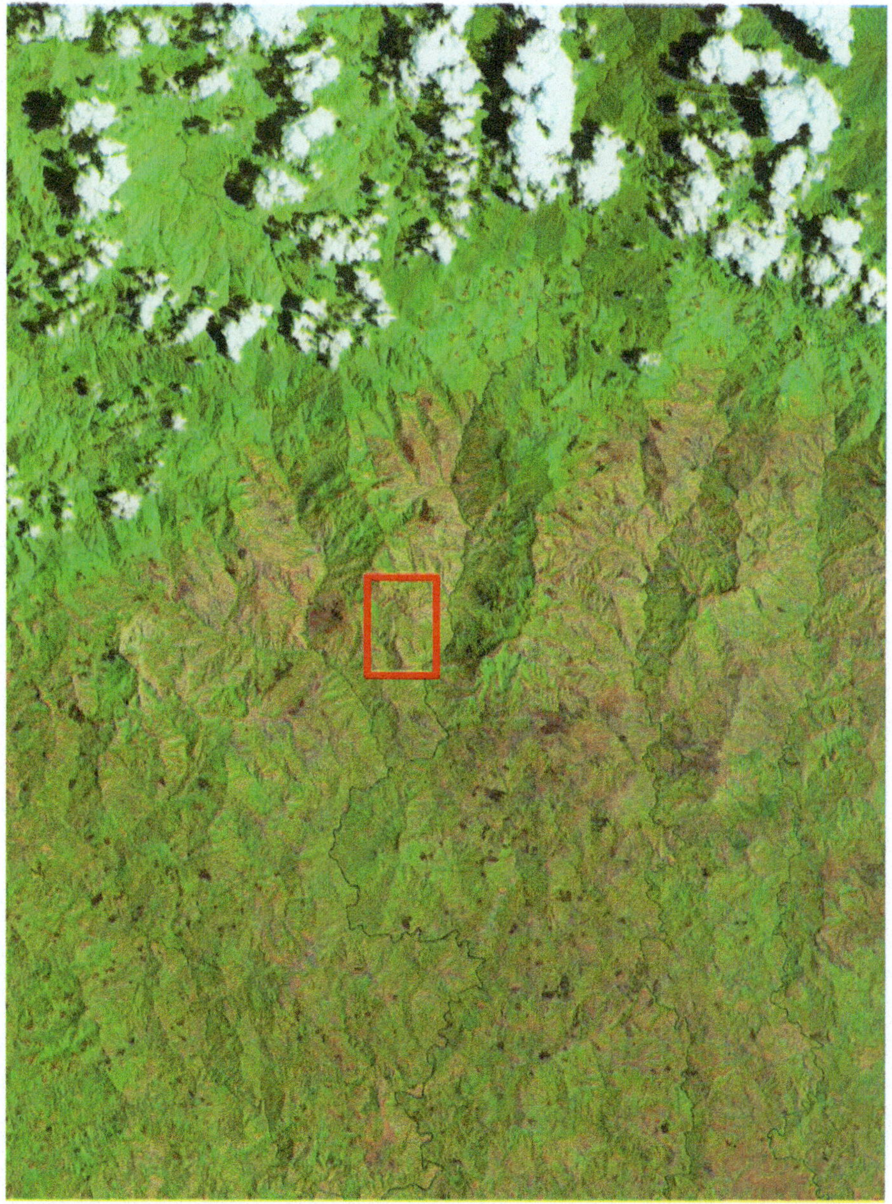

Landsat Image T/M4-5, 26 March 1997

Regional Environment

This site, on the upper Tabasara River in central Panama, is 145 km east of Figure E8 in the same volcanic belt whose forested crest line to the north is at 1,500 m. The 2,800 m high Santiago Volcano is 17 km to northwest in the forest zone.

Physiography

The morphology is characteristic of deeply weathered, strongly dissected tephra. Regional ridges ranging from 1,200 to 1,300 m appear bare. Lower slopes and valleys at about 600 m are vegetated.

E9 Airphoto

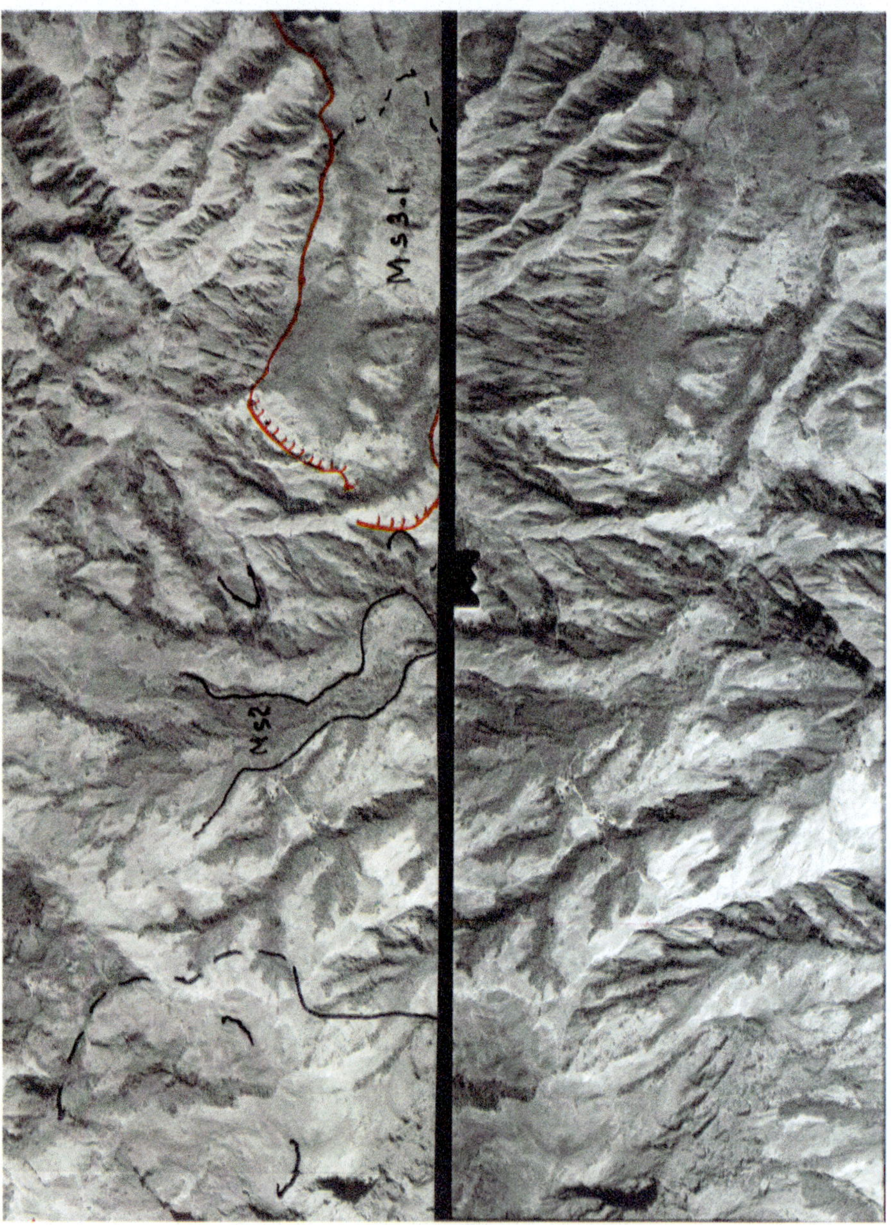

E9 Stereopair

Photo Interpretation

The **Ms3.1** slump in the 1:20,000 scale air photo is 1,800 m long with typical back-tilted blocks on its 600 m wide upper half. The mass of the expanded lower depositional half consists of disintegrated hummocky material. The slide may be at a contact zone of lavas and tephra. The air photos were taken in 1971, recent imagery shows the mass material to be vegetated and partly cultivated in contrast to still bare adjacent terrain. **t** are river terraces. **Ms2** is a small debris slide

Photo Source

Personal archive

Figure E10: Col de Crous

Location 44°08′57″N, 0°53′56″E

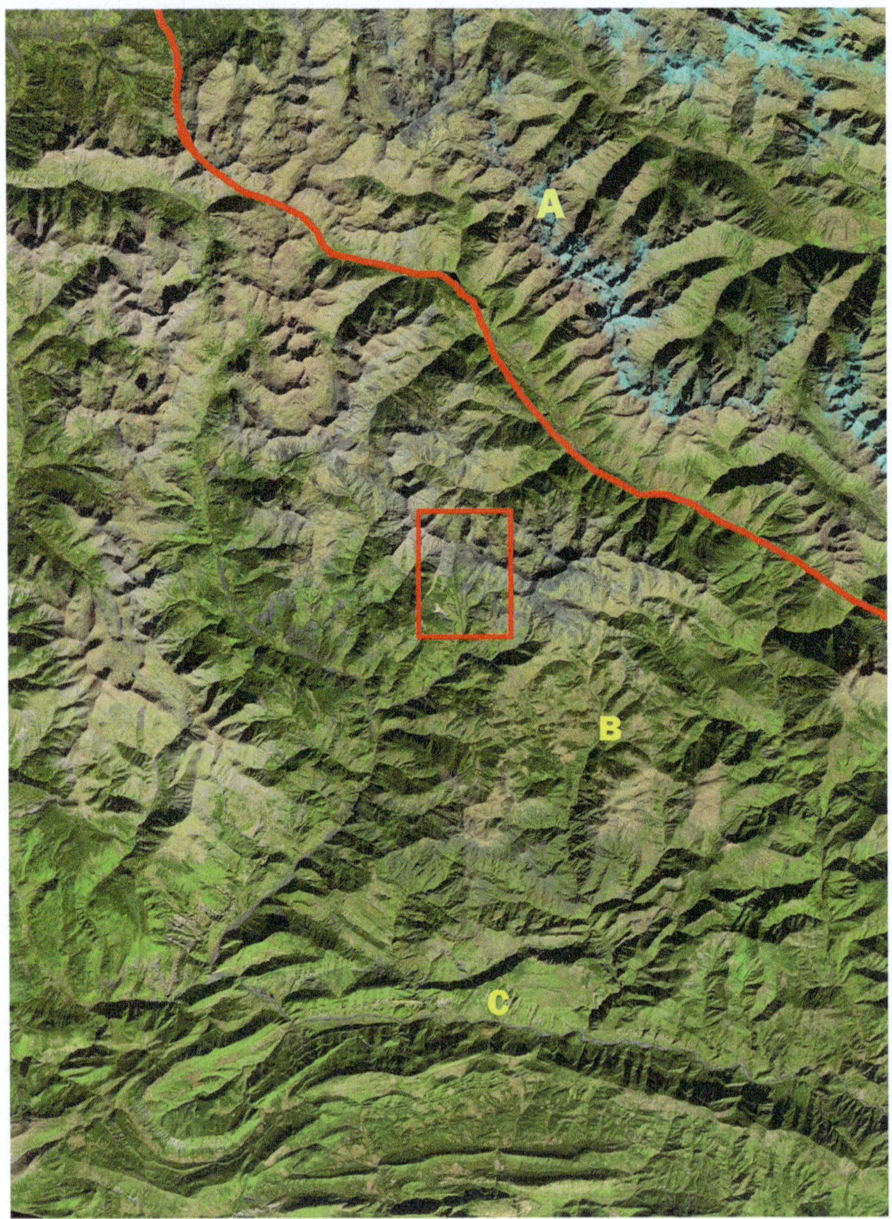

Landsat Image T/M 4-5, 16 October 2003

Regional Environment

The scene covers 1,600 km^2. The example site is in a tectonic zone of the French Maritime Alps.

Structure

The area is confined between the **A** Hercynian Mercantour massif, and the Permian **B** Barrot Dome on the south. The folds at **C** bottom are Cretaceous sediments of the Castellane Arc.

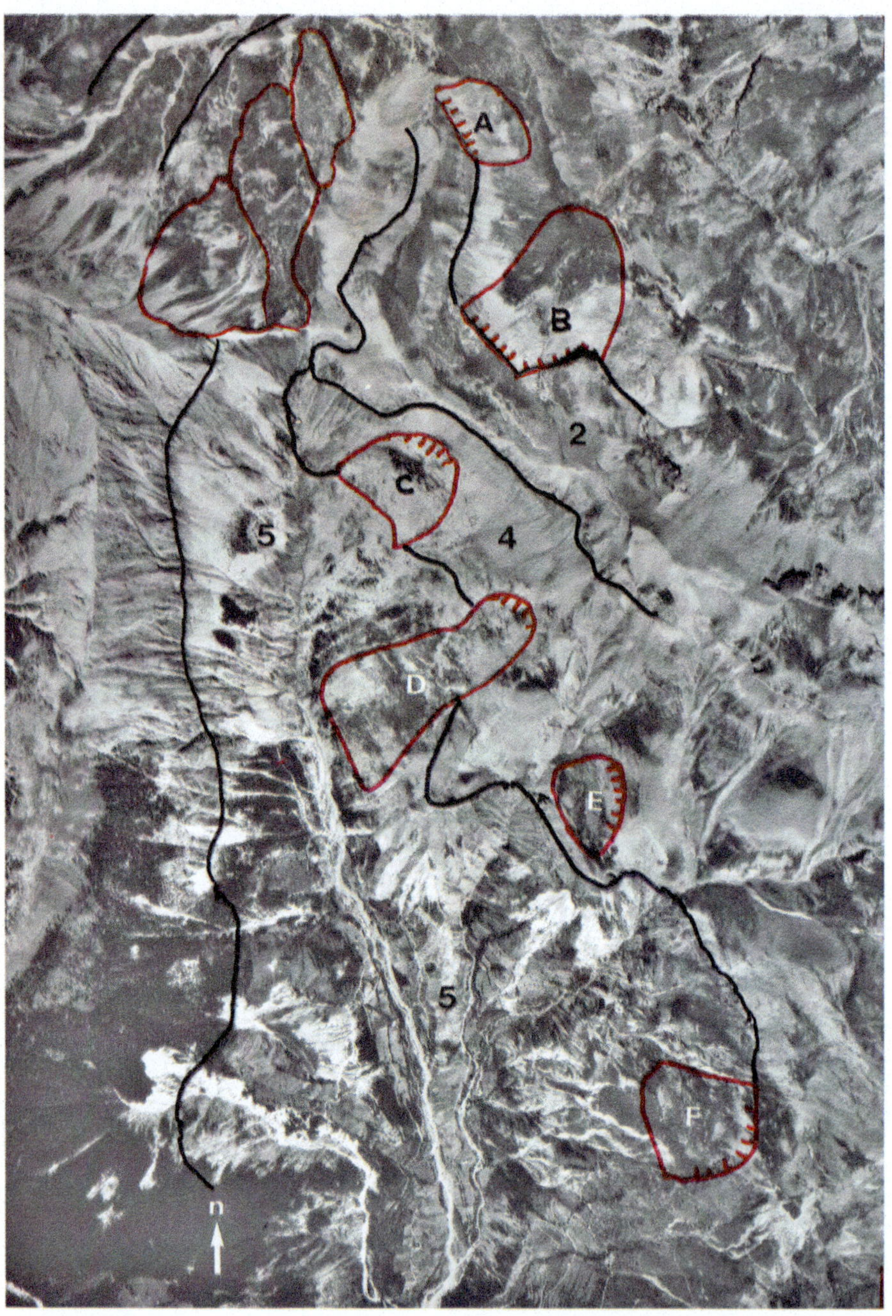

E10 Airphoto

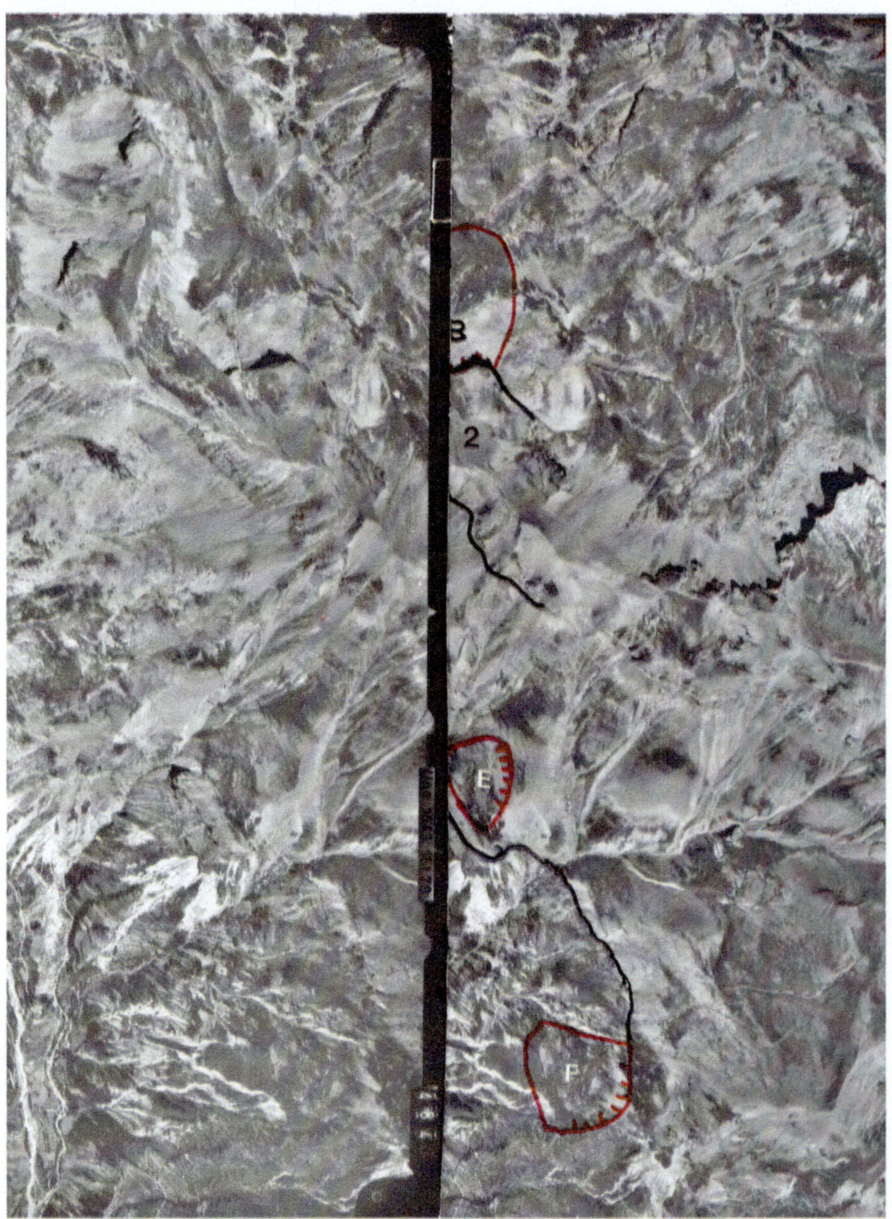

E10 Stereopair

Photo Interpretation

Six occurrences of two movements and three bedrock units are delineated on this 1:25,000 air photo.

Mass Movements
Units A and **B** are debris slides
Units C, D, E, F are rock slumps in weak shaly limestones and marls

Bedrock Units
Units 2 and 4 are ridge makers, Unit 5 is on lower valley slopes
2—is Upper Jurassic limestone, dolomite and marl
4—is partly talus covered Lower Cretaceous shaly limestone and marl
5—is Mid Cretaceous limestone and marl

Photo Source

Figure E11: Bielle

Location 43°03′14″N, 0°30′49″W

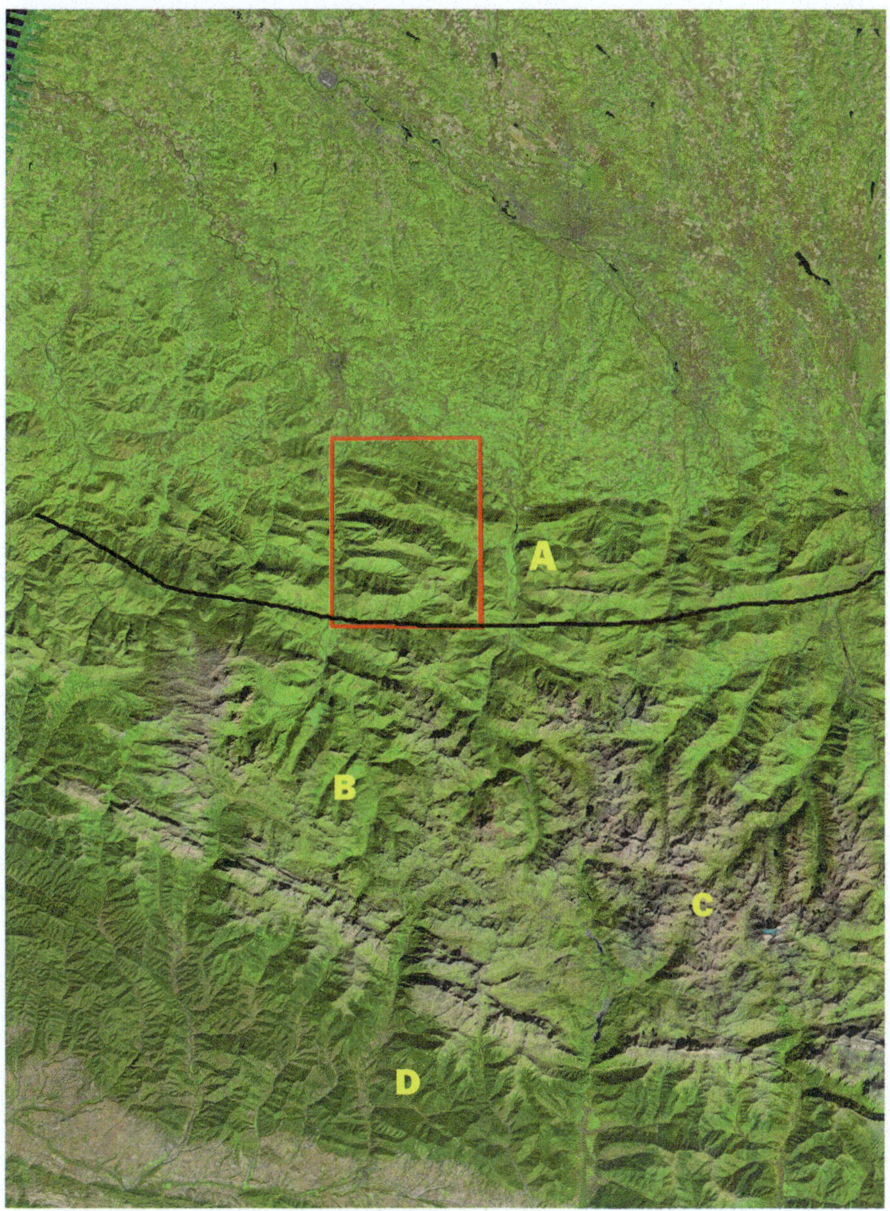

Landsat Image 4-5 TM, 02 October 2011

Regional Environment

The scene covers 5,800 km^2 of the southern Aquitaine Basin and the western Pyrenée Orogen.

Structure

The flat appearing terrain of the north part of the scene is the Miocene upper plain of the Basin.

A—are the Front Ranges

B—are Upper Paleozoic sediments of the Axial Zone

C—is a Hercynian granite stock

D—are Lower Tertiary conglomerate folds of the Spanish Pre-Pyrenees

Physiography

The site is 10 km west of Figure E19. The image shows a succession of steeply dipping anticlinal limestone ridges at elevations ranging from 1,600 to 1,900 m, and intervening shale valleys.

E11 Airphoto

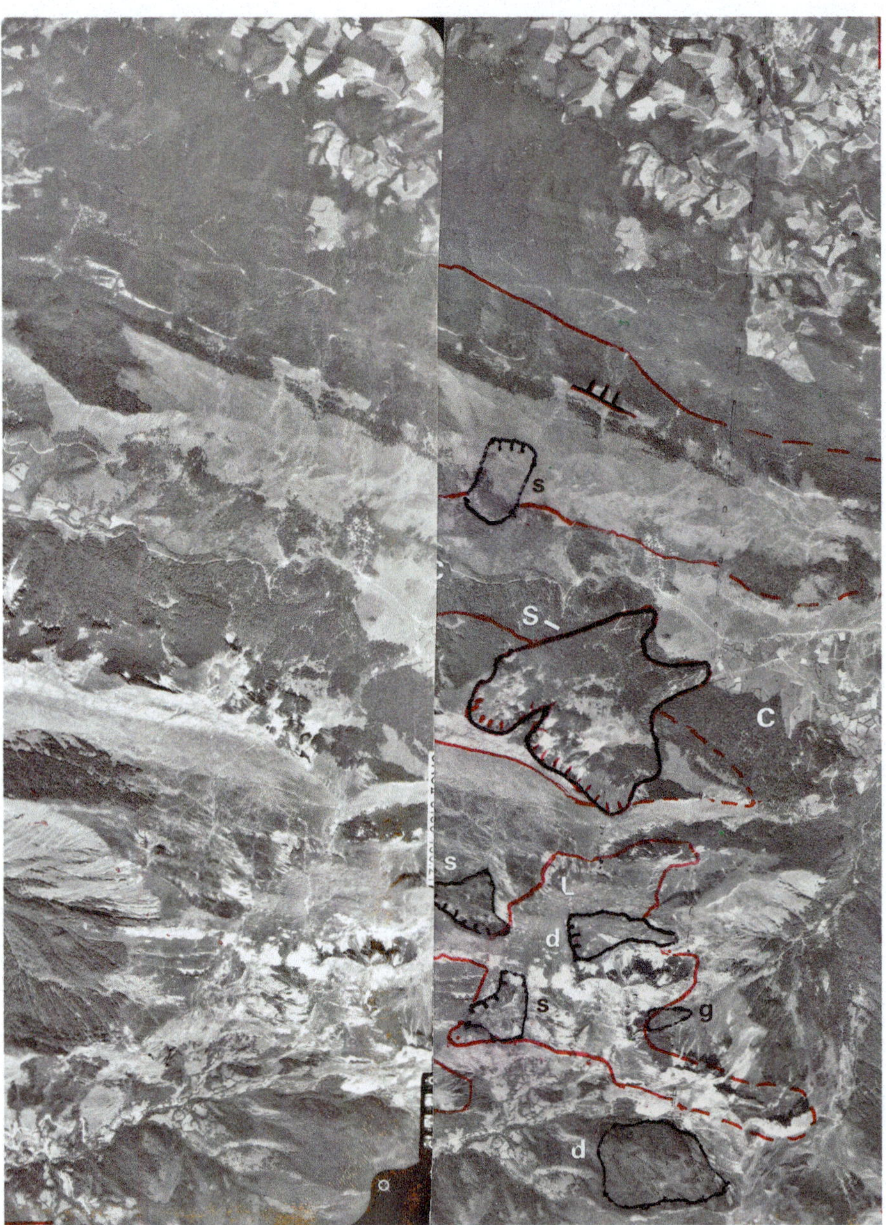

E11 Stereopair

Photo Interpretation

Three types of mass movement and two bedrock types are delineated on this 1:60,000 scale air photo.

Mass Movements

S—are two **rock slumps** on the north slope of a limestone ridge, and two smaller slumps on the crest of another ridge

D—are **debris slides**

G—is a small **rock glacier**

Bedrock Units

L—are dominant Mid-Jurassic limestone ridges

C—are Lower Cretaceous shales

Photo Source

Copyright IGN, 1979, 600, 160-161

Figure E12: Ramboda

Location 07°02′N, 80°43′E

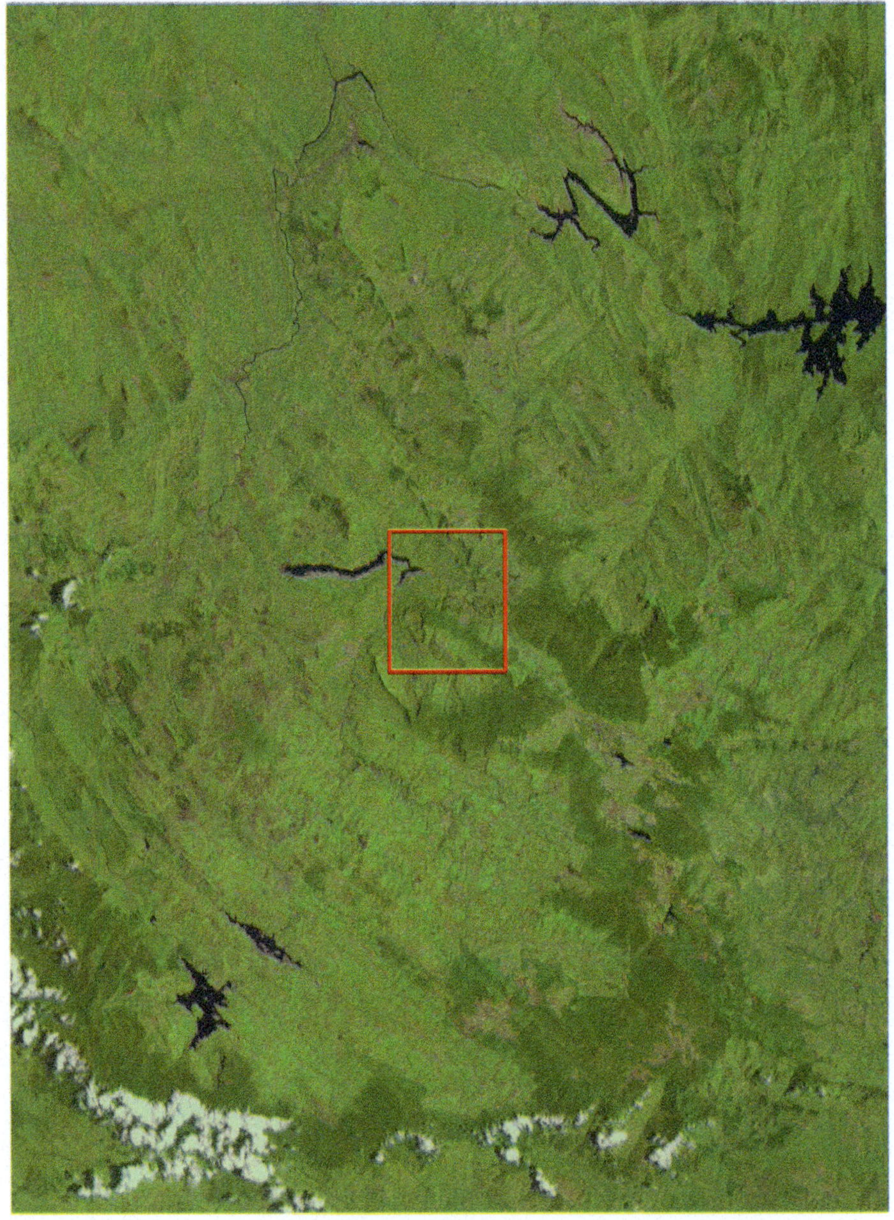

Landsat Image L7 S2C-on, 14 March 2001

Regional Environment

The site is 20 km north of Figure B3.

The structural divisions and physiography of the area are described in Figure B3.

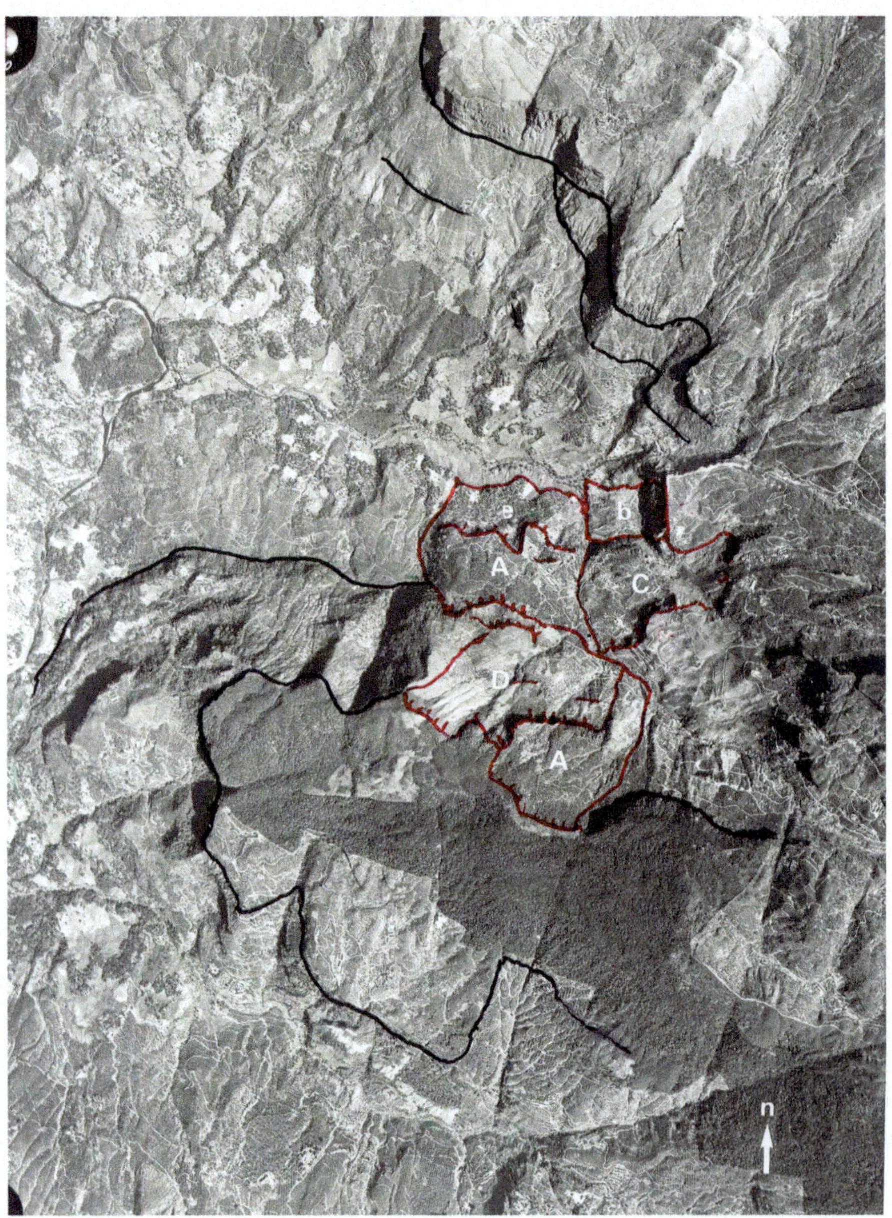

E12 Airphoto

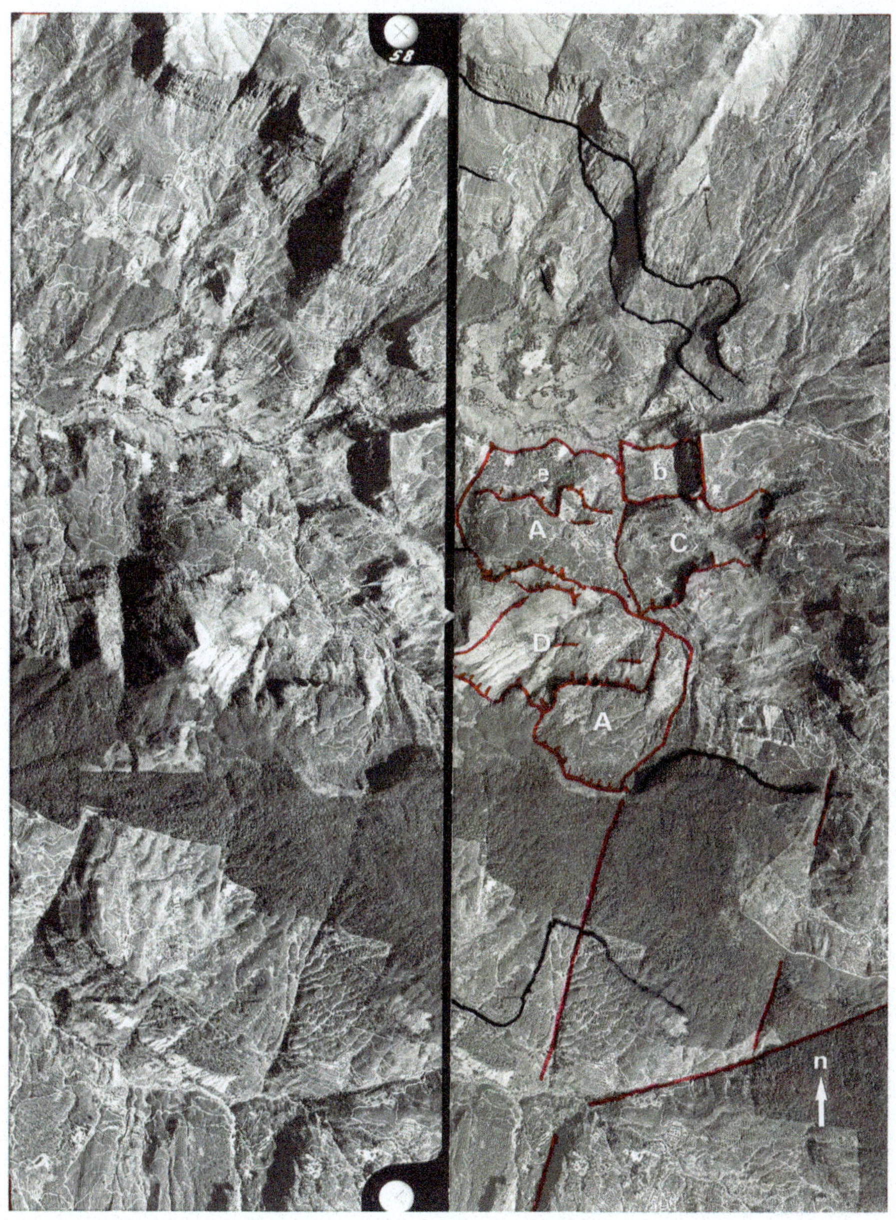

E12 Stereopair

Photo Interpretation

The black-traced escarpments on the air photo are related to extensive normal faulting and fracturing in the basement rocks of Sri Lanka in the Mesozoic Era. The weaker valley rocks are metasediments. The main forested area is the axial ridge of an interbedded anticline of granite and metasediments. Four types of mass movement occur on the upper slopes of the valley ridges in the 1:40,000 scale air photo.

A—rock slumps are identified at two levels on the ridge; one at the ridge crest and a second at a lower stratum bench

B and **e**—are two **debris slides,** one from a bench scarp and the other below the lower rock slump

C—is a larger **debris avalanche** between the rock slump and the debris slide

D—is large **rock slide** adjoining the rock slump

Photo Source

Personal archive

Figure E13: Villar d'Arène 1

Location 45°02′N, 06°20′E

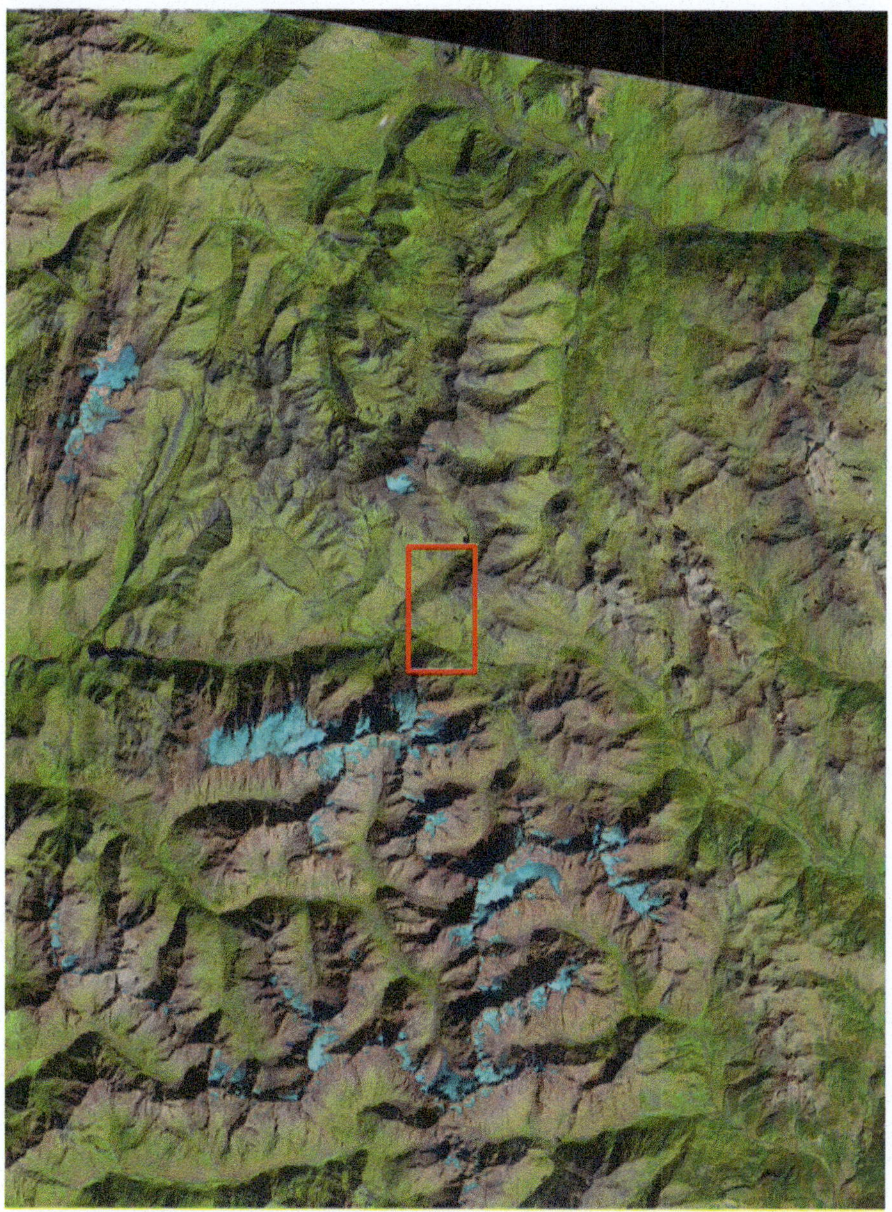

Landsat Image T/M 4-5, 21 September 2003

Regional Environment

The scene covers 1,575 km^2 in the central Alps 50 km east of Grenoble.

Structure

The east-west Romanche valley in the center divides the area into two distinct tectonic terrains.

The north half is in the locally 20 km wide Alpine Furrow of Lower Jurassic interbeds of limestone shale and marl with elevations ranging from 2,000 to 3,000 m.

The south half are crystalline rocks of the north end of the Pelvoux Massif at 3,000 to 4,000 m.

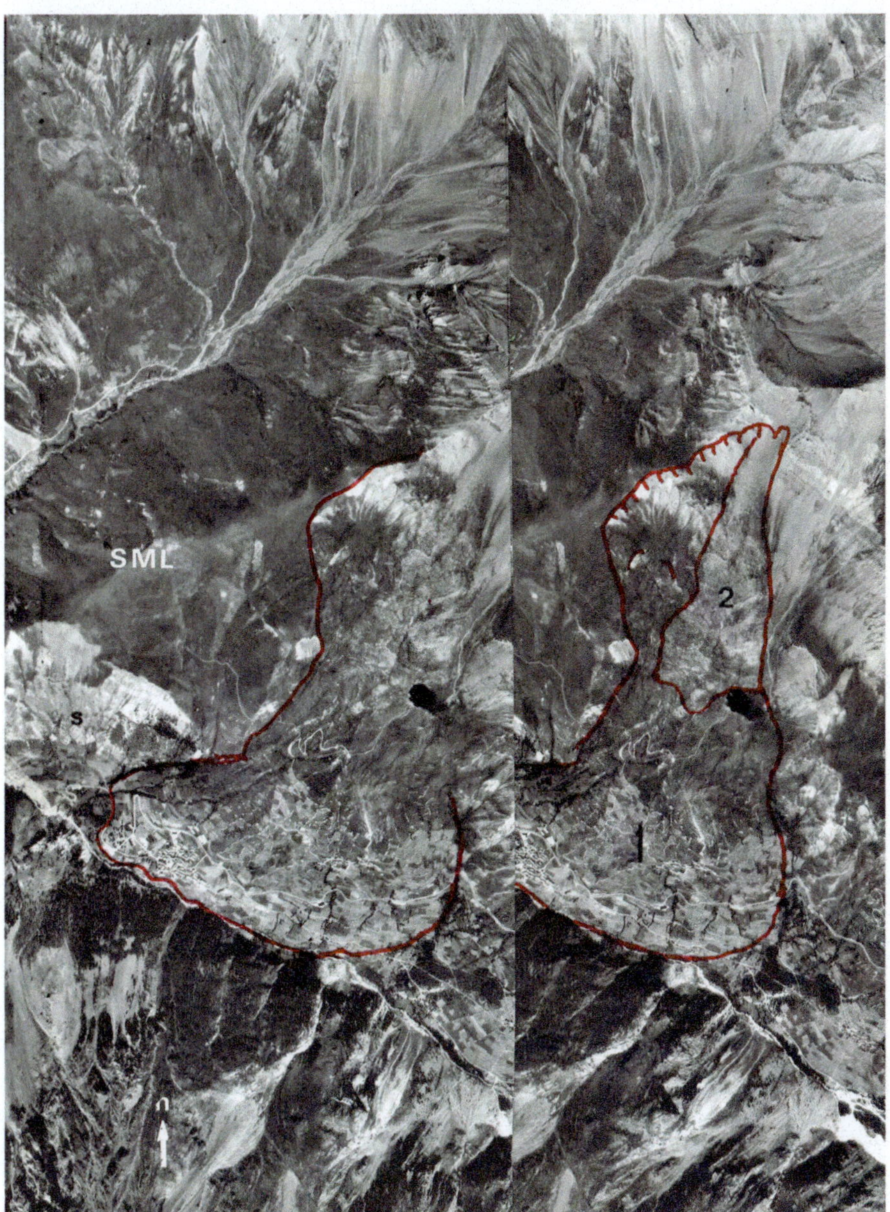

E13 Airphoto

Photo Interpretation

This **stereogram** at scale 1:20,000 delineates **Unit 1** of a major **debris avalanche** in Lower Jurassic interbedded sediments lying against the Pelvoux rocks. The foot of the slide is 1,200 m wide.

2—is a **rock avalanche**

S—is a 700 m wide rock slide in **SML** shale, limestones and marls. The failure is related to the debris avalanche on the opposite bank of the river discussed in Figure E14.

Photo Source

Copyright IGN, 1974, 3435

Figure E14: Villar d'Arène II

Location 45°02′N, 06°20′E

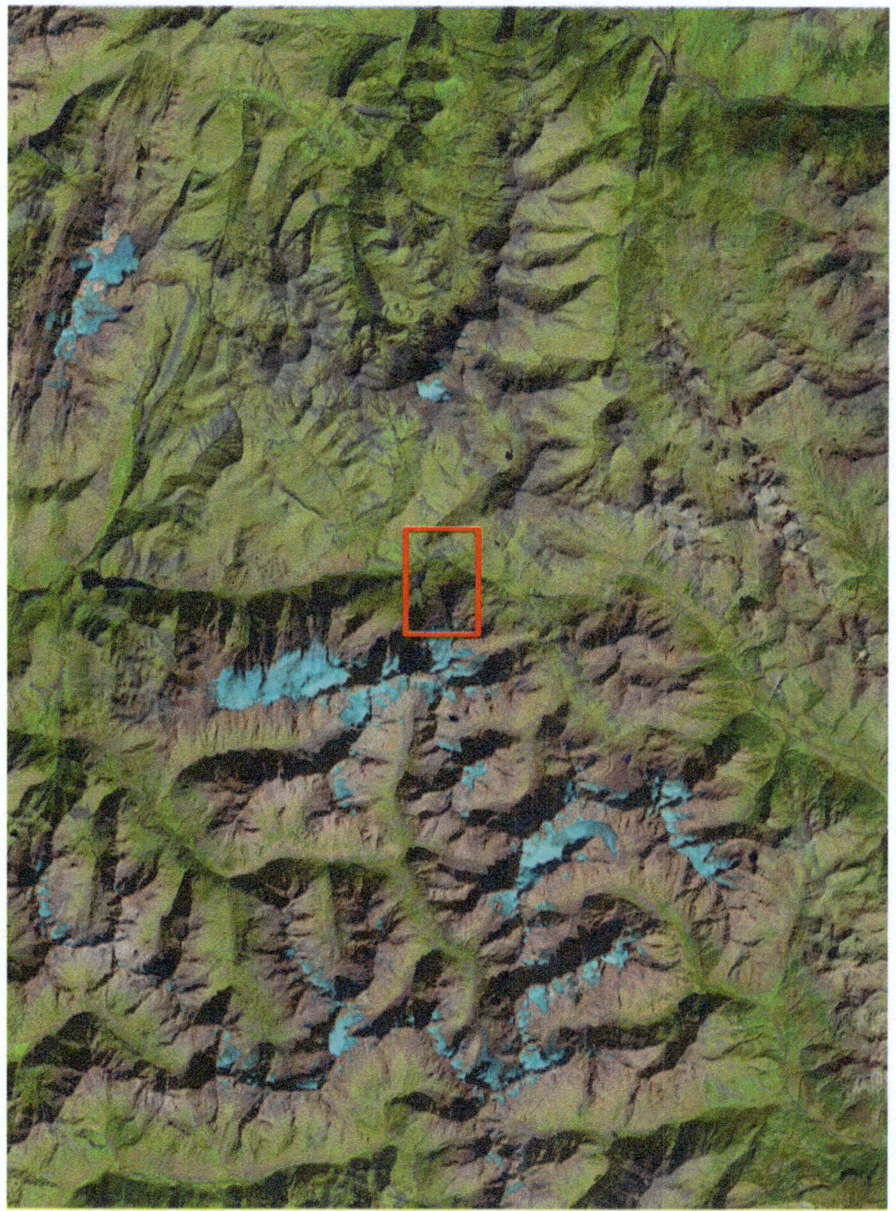

Landsat Image T/M4-5, 21 September 2003

Regional Environment

This is the same image as Figure E13

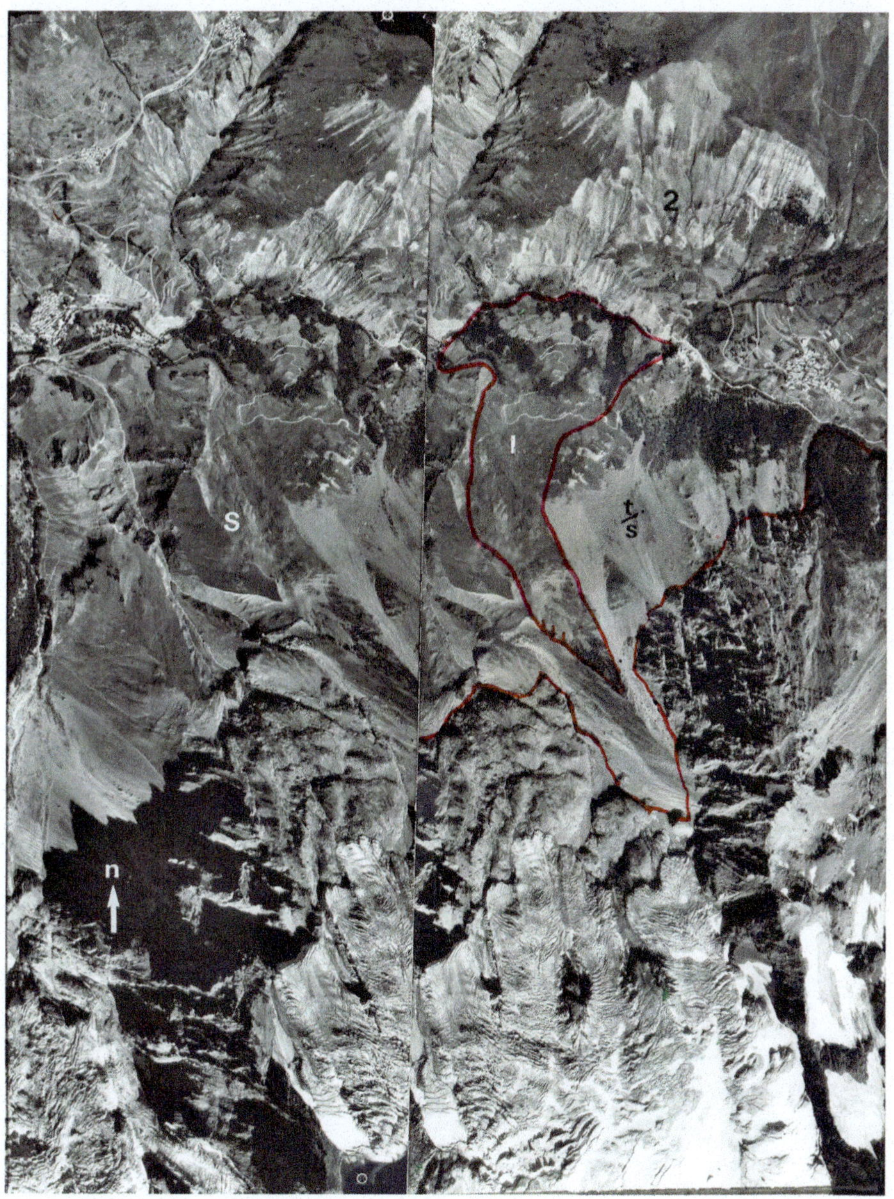

E14 Stereopair

Photo Interpretation

The **stereogram** of this 1; 20,0000 air photo shows four mass movements associated with those on the north side of the Romanche valley.

1—is a 1,500 m long **debris avalanche** which lies on a slope recessed onto the front of the thrusted north Flank of the crystalline massif. It is **S** of the same lithology as the SML rocks of E 13. It has a foot width of 700 m. Its 30 m thick toe mass has deflected the river 200 m to the base of the rock slide **2** in weak shales and marls on the opposite bank. The river has undercut the base of the slide sufficiently so that the valley roadway crossing that point has been protected by a tunnel set back behind the failure. A high risk slope instability warning was issued for this area on 24 June 2011.

t/s is a 500 m wide talus slope.

Photo Source

Copyright IGN, 1974, 3435, 81-82

Figure E15: Cima Argentera

Location 44°09′N, 07°22′E

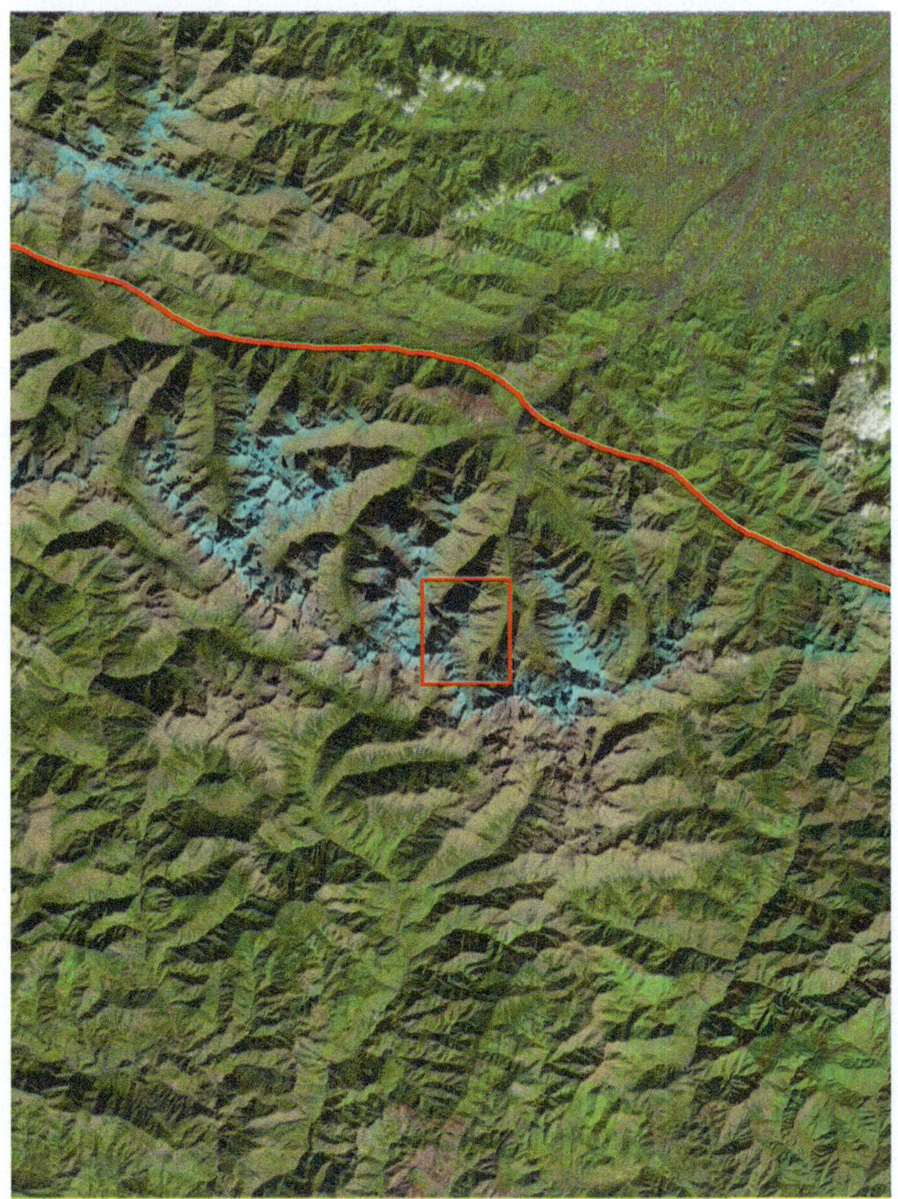

Landsat Image T/M 4-5, 16 October 2003

Regional Environment

The image shows the Example site to be near the center of the Hercynian Mercantour Massif with snow-covered peaks. The site is 4 km inside the Italian border. The smooth appearing area in the northeast is the westernmost plain of Italy's upper Piedmont at Cuneo. The parallel ranges between the Plain and the Massif are Jurassic and Cretaceous folds of sedimentary rocks.

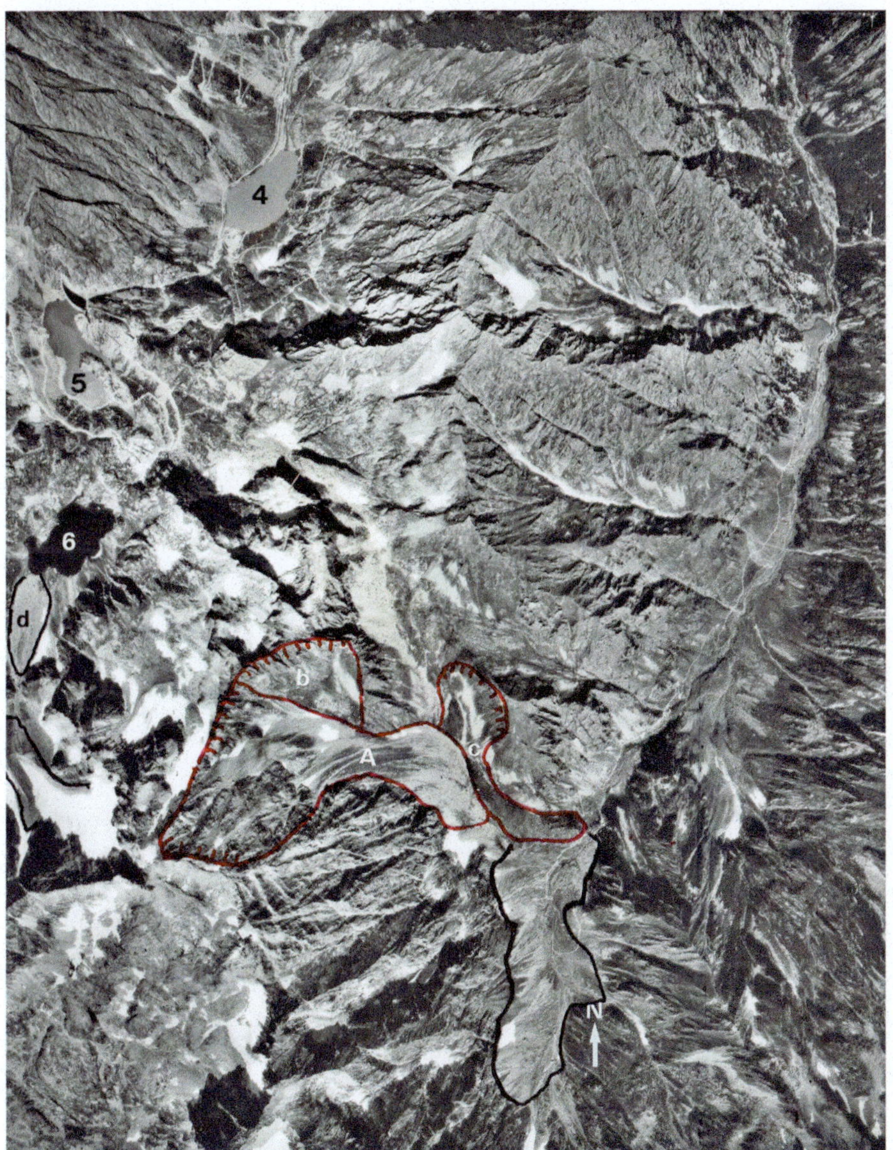

E15 Airphoto

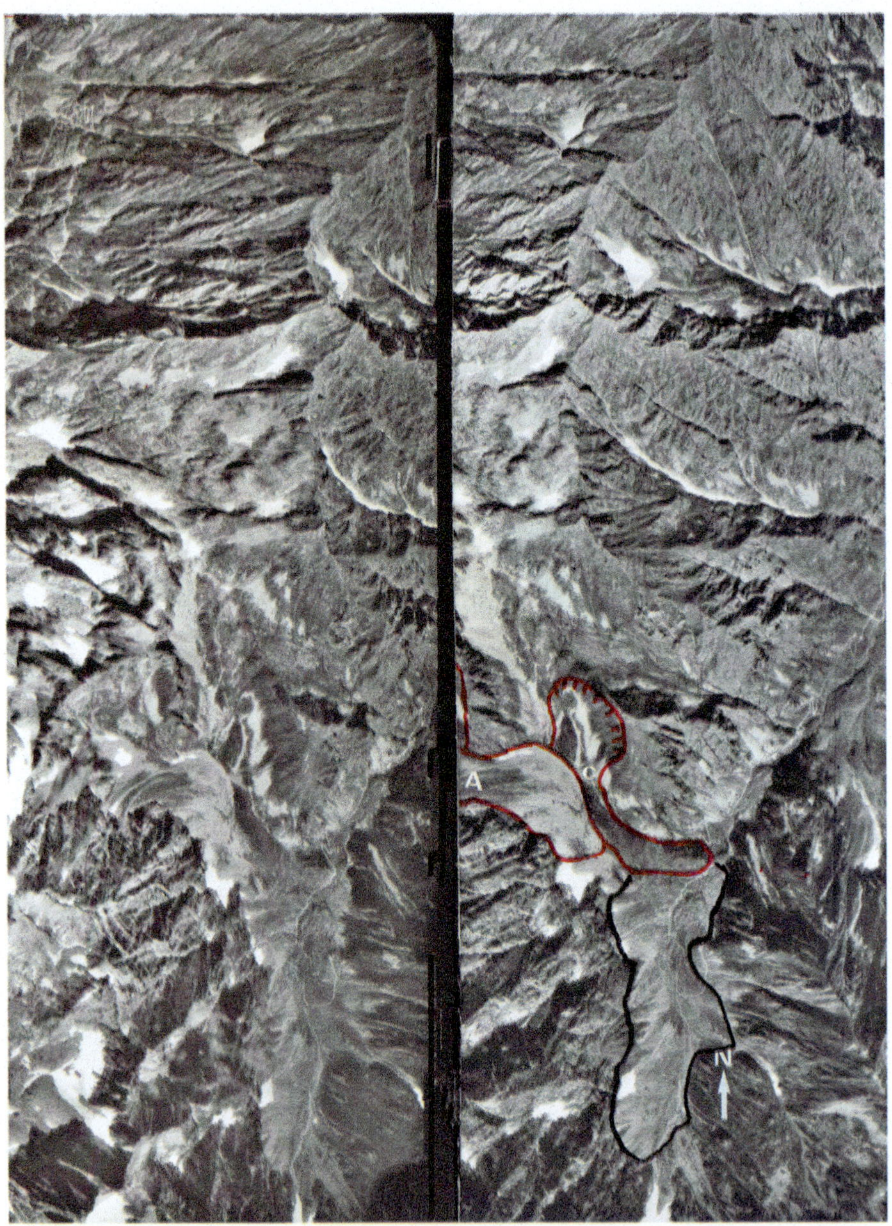

E15 Stereopair

Photo Interpretation

This 1:30,000 air photo shows a group of three mass movements in steep gullies.

A—is an active appearing debris avalanche which came to rest midway down the gully

b—is a debris fall below a vertical cliff

c—is a slide of dark debris that flowed down into the stream gorge. It originated in a local rock source of different composition from the main mass.

Lakes numbered **4**, **5** and **6** northwest of the landslide group are linked hydroelectric power reservoirs. A zone of alluvial fans are traced on the lower valley slopes south of the slide group.

Photo Source

Copyright IGN, 1978, 3540, 93-94

Figure E16: Fournel

Location 44°48′16″N, 06°25′28″E

Landsat Image T/M 4-5, 16 October, 2003

Regional Environment

Structure

The image covers 3,500 km^2. It includes the snow-covered Pelvoux Hercynian Massif in the north and the flysch nappes on the east of the winding Durance valley and south to the Embrun Reservoir. The area northwest of the site is the most southerly glacial massif of the Alps.

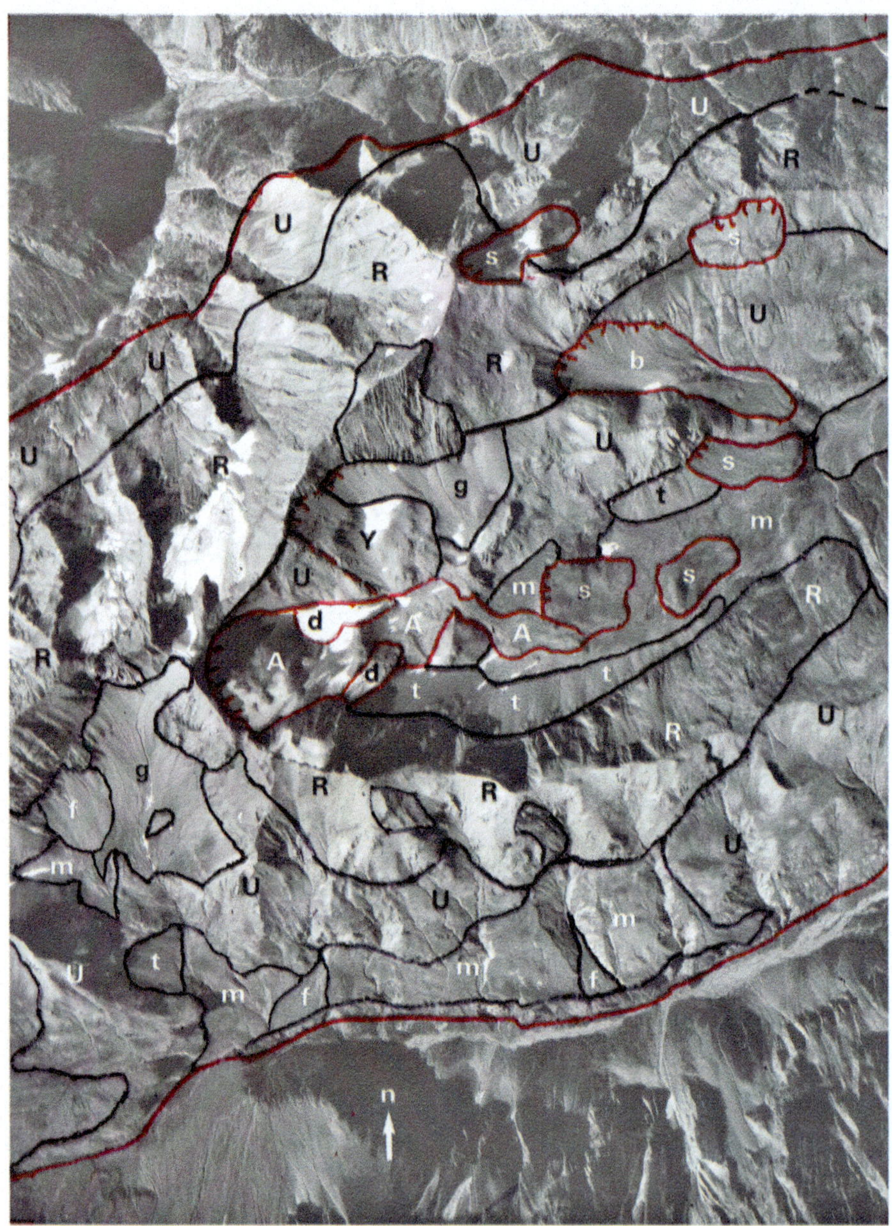

E16 Airphoto

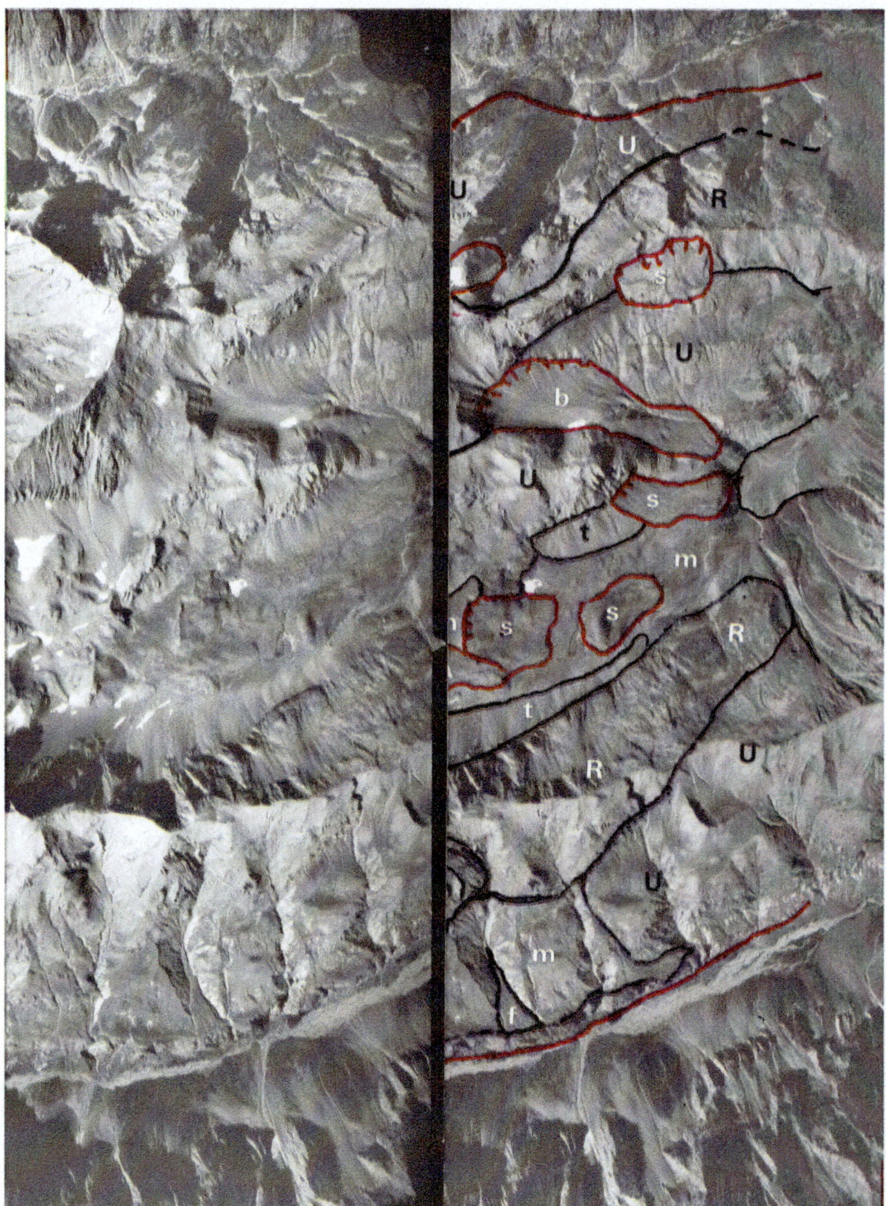

E16 Stereopair

Photo Interpretation

The outer marginal red lines on the air photo mark the extent of part of a 20 km^2 anticline unit of interbedded Jurassic sedimentary rocks lying as a nonconformable inlier on the Hercynian granites. The slides are associated with the contacts of beds of varying resistance. The site is 10 km east of E17. Eleven geounits of five categories are delineated on this 1:30,000 scale air photo.

Mass Movements

These occur mainly in the shale core of the fold:

A is a 1.5 km long **debris avalanche** which originates at 3,100 m at the high point of the fold axis

b is a kilometer long **debris flow**

S are three **debris slides**

Y is a **rock slide**

g are **debris flows**

t are **talus** slopes

Fluvial Surficial Deposits

f are small **alluvial fans**

Glacial deposits

m are **moraine** deposits along the stream valley

Periglacial Forms

D are two **rock glaciers** on the margins of the debris avalanche

Bedrock Units

R are **Lower Jurassic limestone** ridges that make up the limbs of the anticline

U are **Lower Jurassic shale** that occur in the core and on the lower limbs of the fold

Photo Source

Copyright IGN, 1970, 3437, 91-92

Figure E17: Hautes Severaisses

Location 44°47′10″N, 06°17′07″E

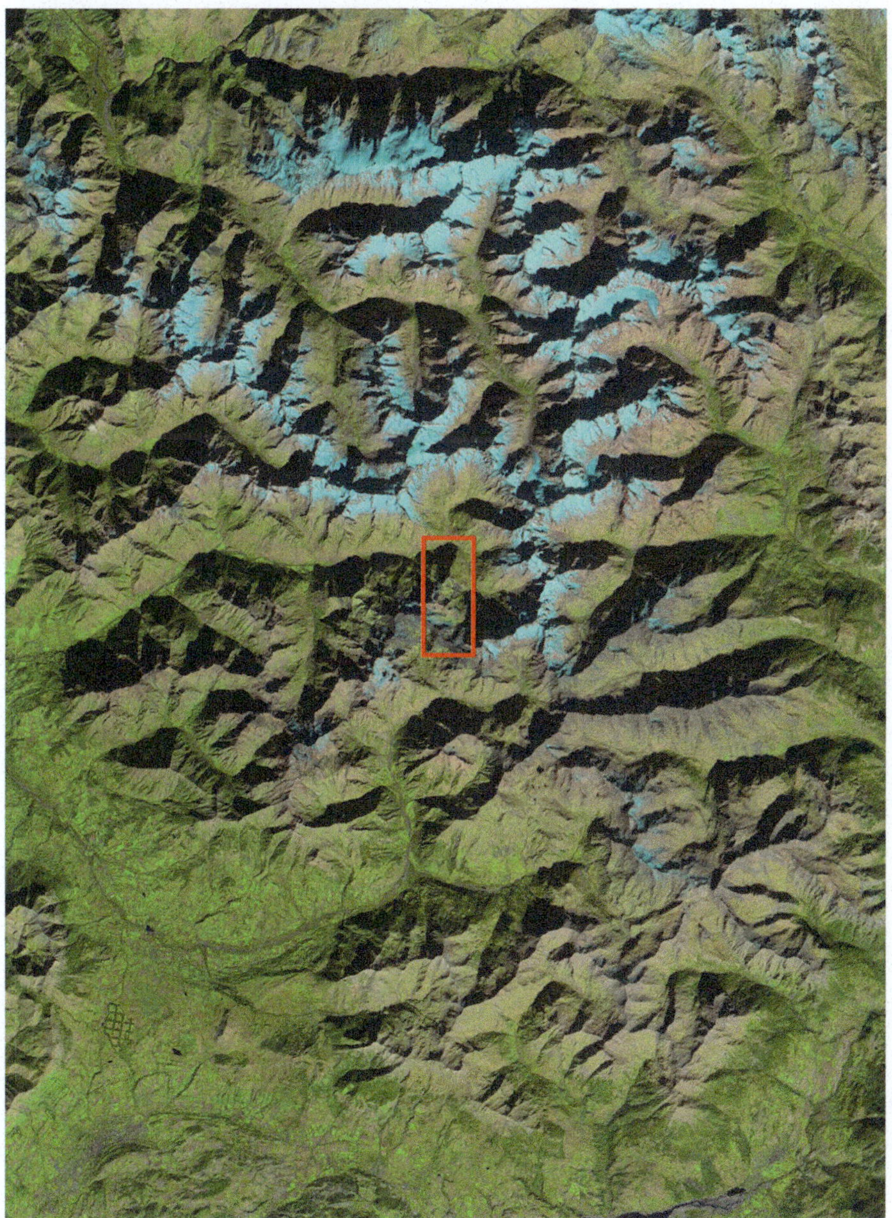

Landsat Image T/M 4-5, 16 October 2003

Regional Environment

The multiple sites visible on the air photo are 10 km west of E16 in the Hercynian massif.

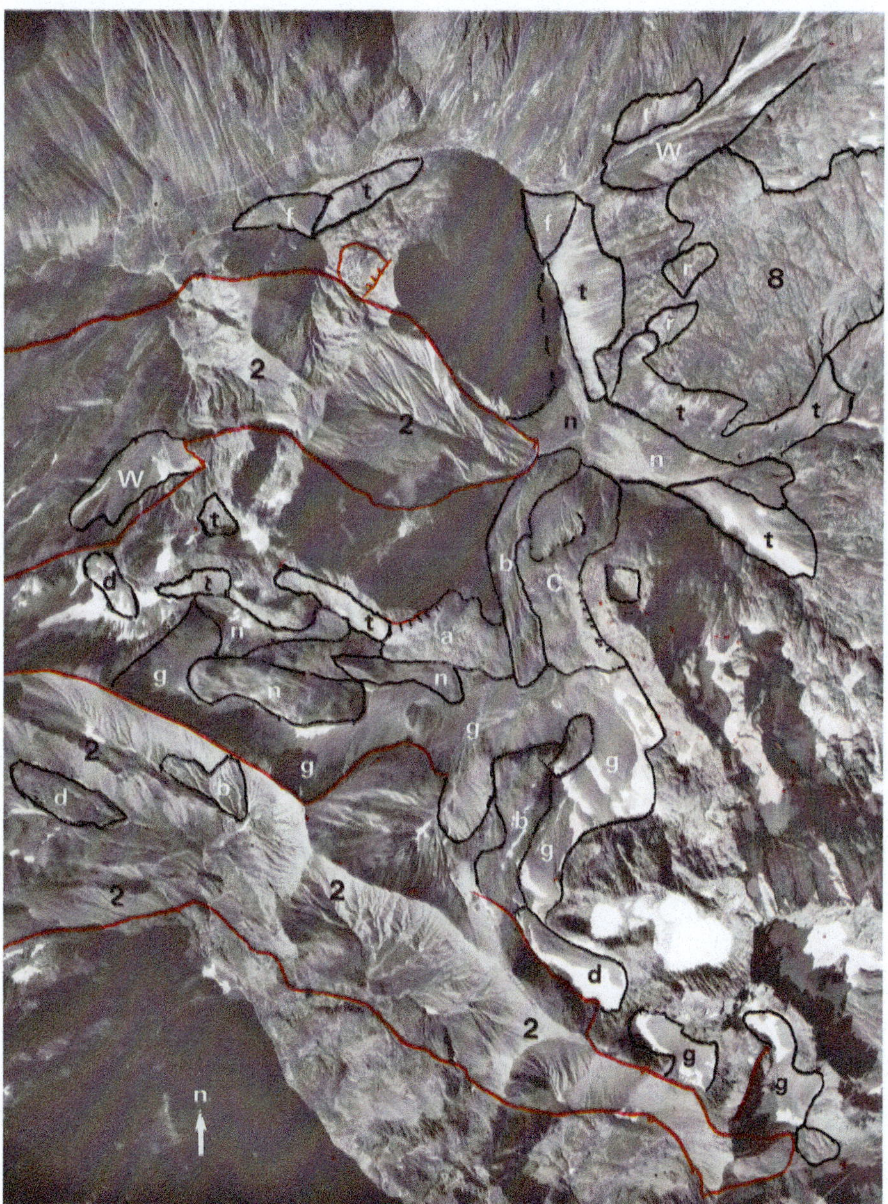

E17 Airphoto

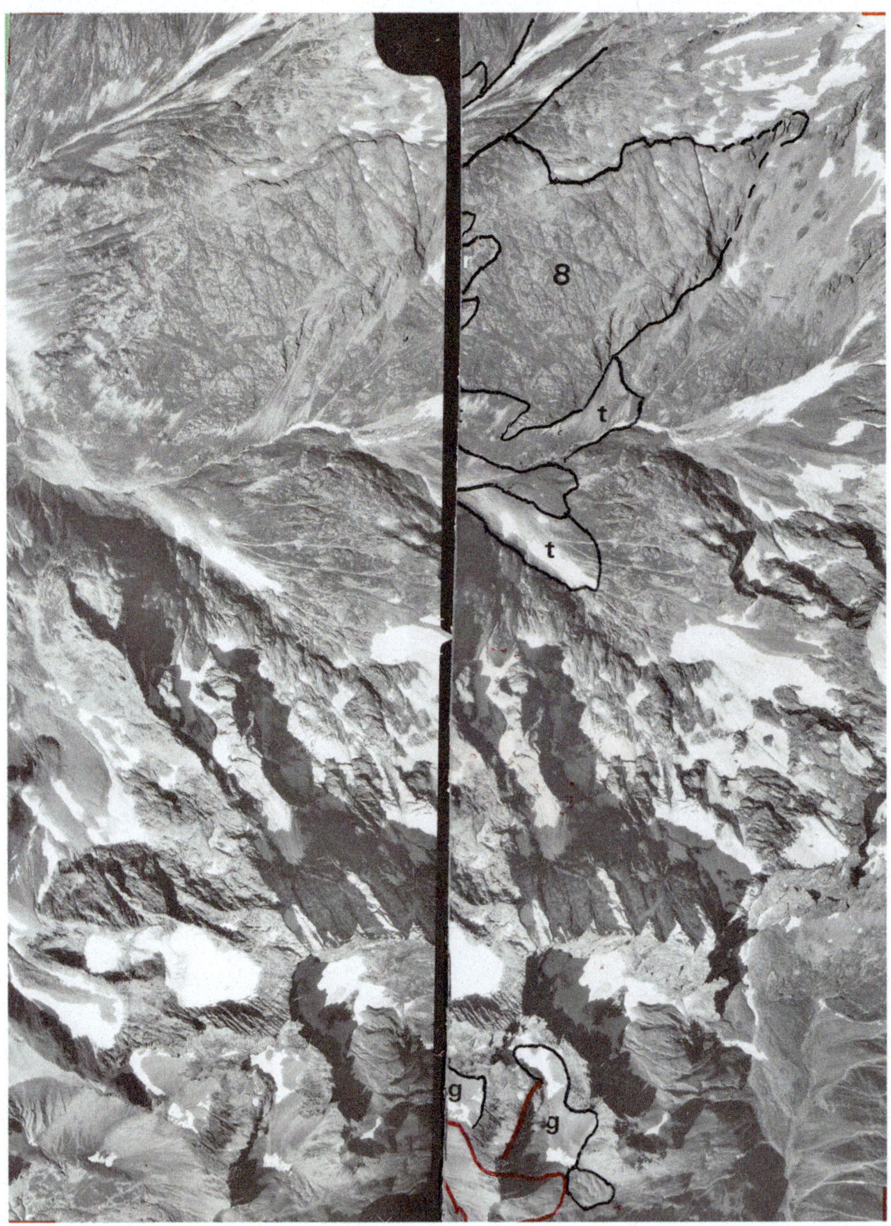

E17 Stereopair

Photo Interpretation

Geounits of five categories occur in this 1:30,000 air photo.

Mass Movements
a is a 600 m wide **rock avalanche**
b is a 1.5 km long **debris slide**
C is a **debris avalanche** adjacent to the slide
W are **debris flows**
r are small **rock slides** on the granite
t are numerous **talus slopes**

Glacial Deposit
n are three relict **marginal moraines**

Periglacial Deposits
d are small **rock glaciers**
g are extensive **gelifluction sheets**

Surficial Deposits
f are small **alluvial fans**

Bedrock Units
2 are Lower Jurassic **lutites**
8 is a slab-like mass of Permian **granite** resting on the country rock
Country rock these are **gneisses and migmatites** of undifferentiated Paleozoic age

Photo Source

Figure E18: Marsous

Location 42°58′06″N, 0°13′15″W

Landsat Image T/M 4-5, 02 October 2011

Regional Environment

The image covers essentially the same area as that of Figure E11.

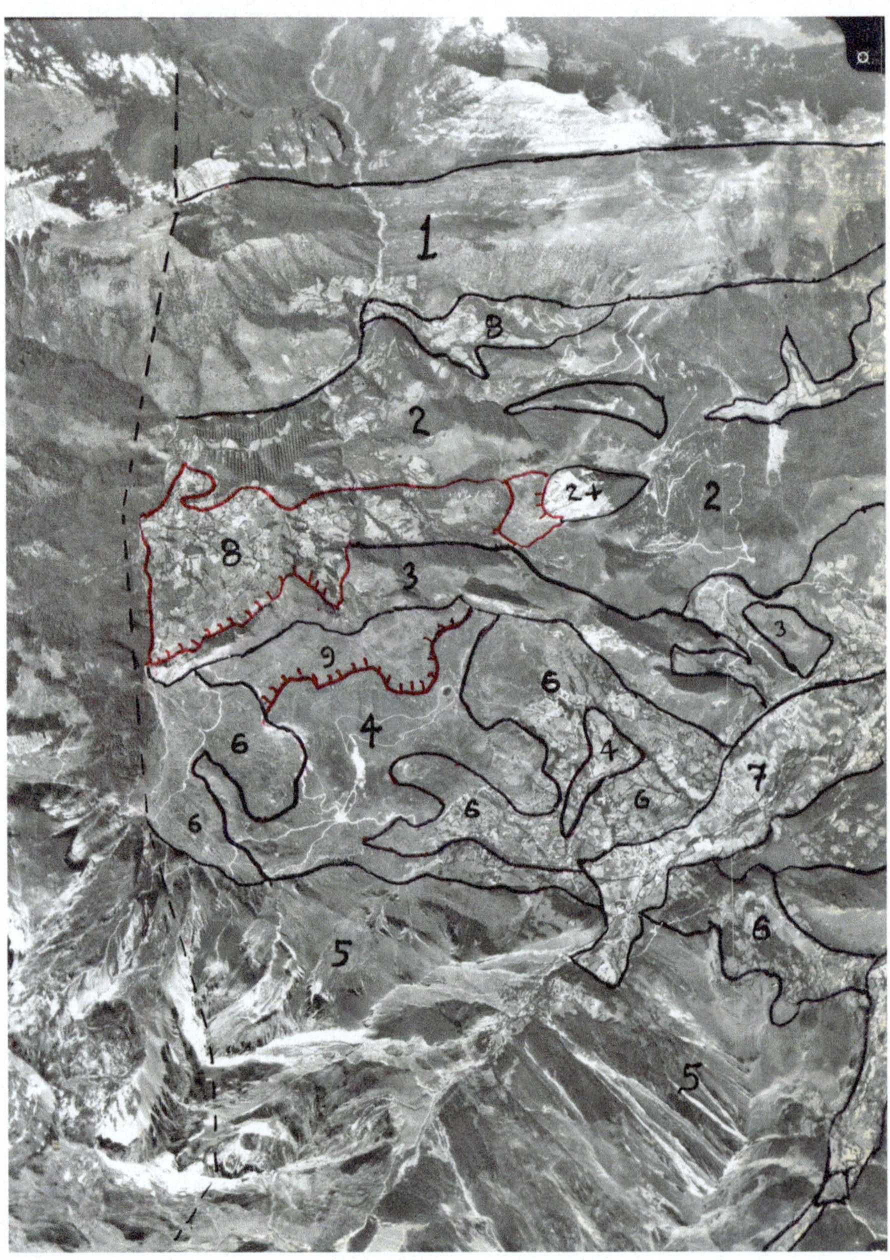

E18 Airphoto

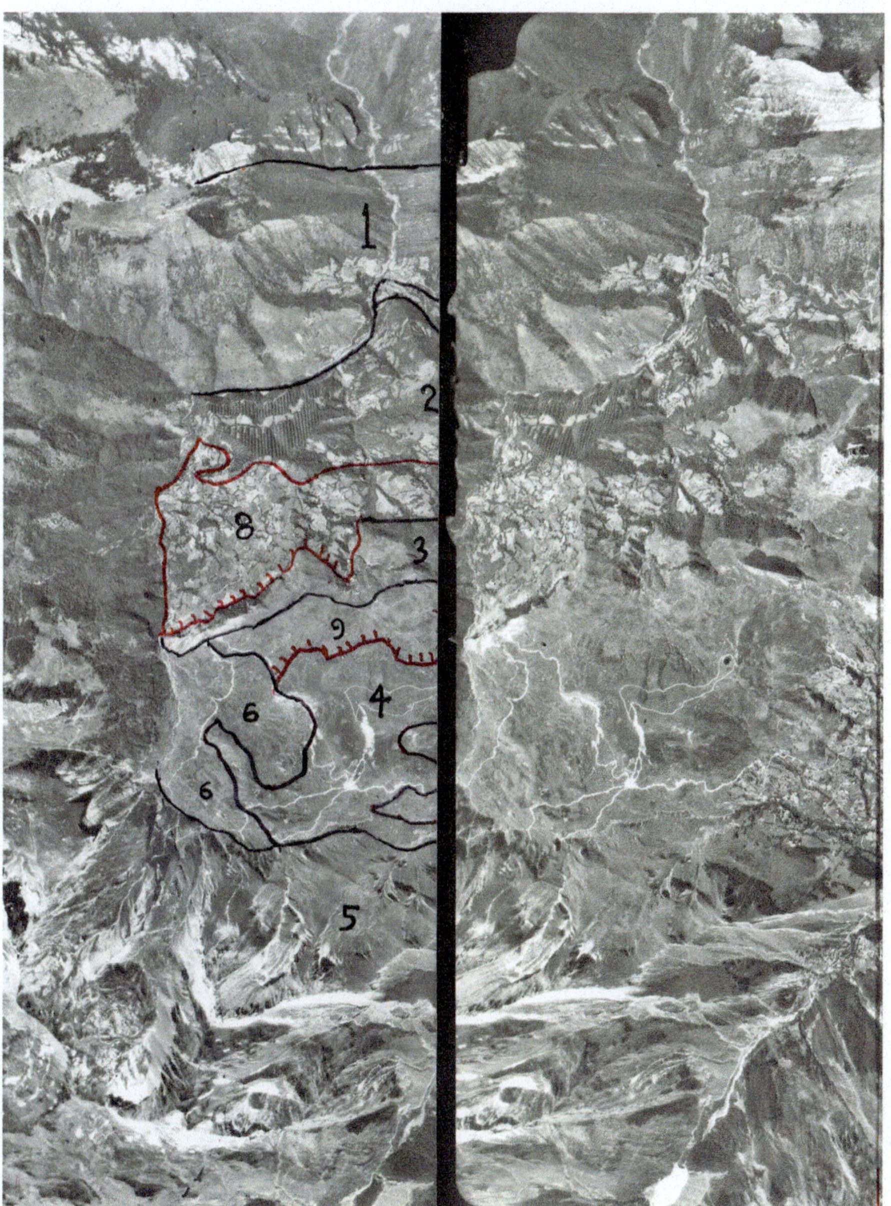

E18 Stereopair

Photo Interpretation

The air photo is 1:60,000 scale.

The site is an extension eastward of E11 in the western Pyrenée Orogen. The area covers a segment of the front ranges and the interior sierras.

1—is a belt of isoclinally folded Mid Jurassic and Cretaceous calcareous rocks ranging from 1,600 to 1,800 m in elevation

2—are a subparallel belt of south-dipping Lower Devonian limestones and shales at 1,350–1,400 m elevation. **2+** is a bare 1,800 m elevation limestone peak

3—is a band of narrow ridges folded Mid Devonian calcareous rocks

4—is an anticlinal sequence of Lower Devonian shales and sandstones at 1,450–1,500 m elevation

5—are high south thrusting interbeds of Lower Devonian limestones and shales at 2,200–2,700 m elevation

6—are cultivated deposits of Mid Pleistocene ground moraine at 1,400–1,000 m elevation

7—is an also cultivated glaciolacustrine valley at 800 m elevation.

8—is an extensive area of debris slides in Unit **2** sediments

9—is an area of debris slides on the north side Unit **4** sediments

Photo Source

Copyright IGN, 1979, 600, 13-14

Reference

Notice explicative de la carte géologique de la France au 1:50,000 Argelès – Gazost 2003, 46 p

Figure E19: Lemieux

Location 45°22′N, 75°06′W

Landsat Image T/M 4-5, 09 October 2011

Regional Environment

The scene covers 320 km^2 of cultivated pattern of glaciomarine clays and green forested deltaic sands of Champlain Sea sediments of eastern Ontario described in Figure E21.

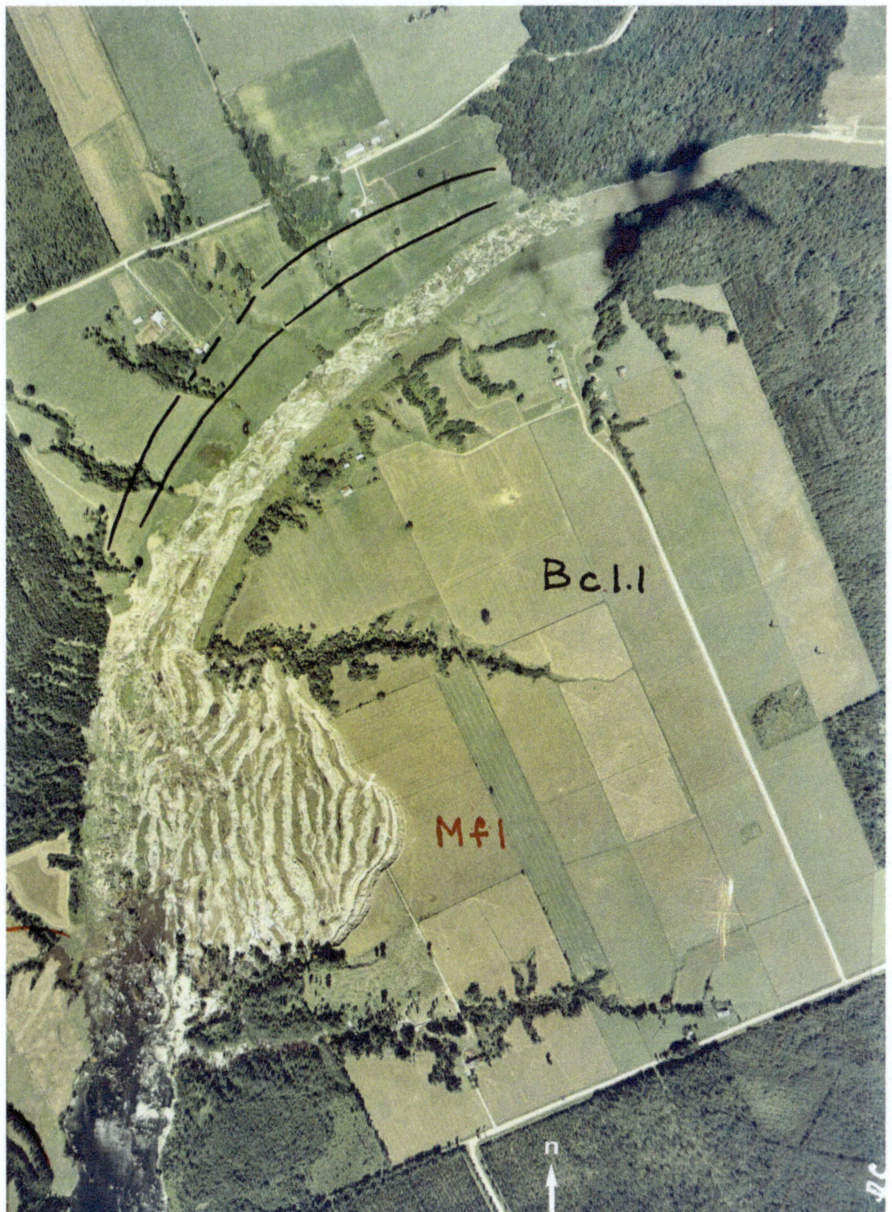

E19 Airphoto (A)

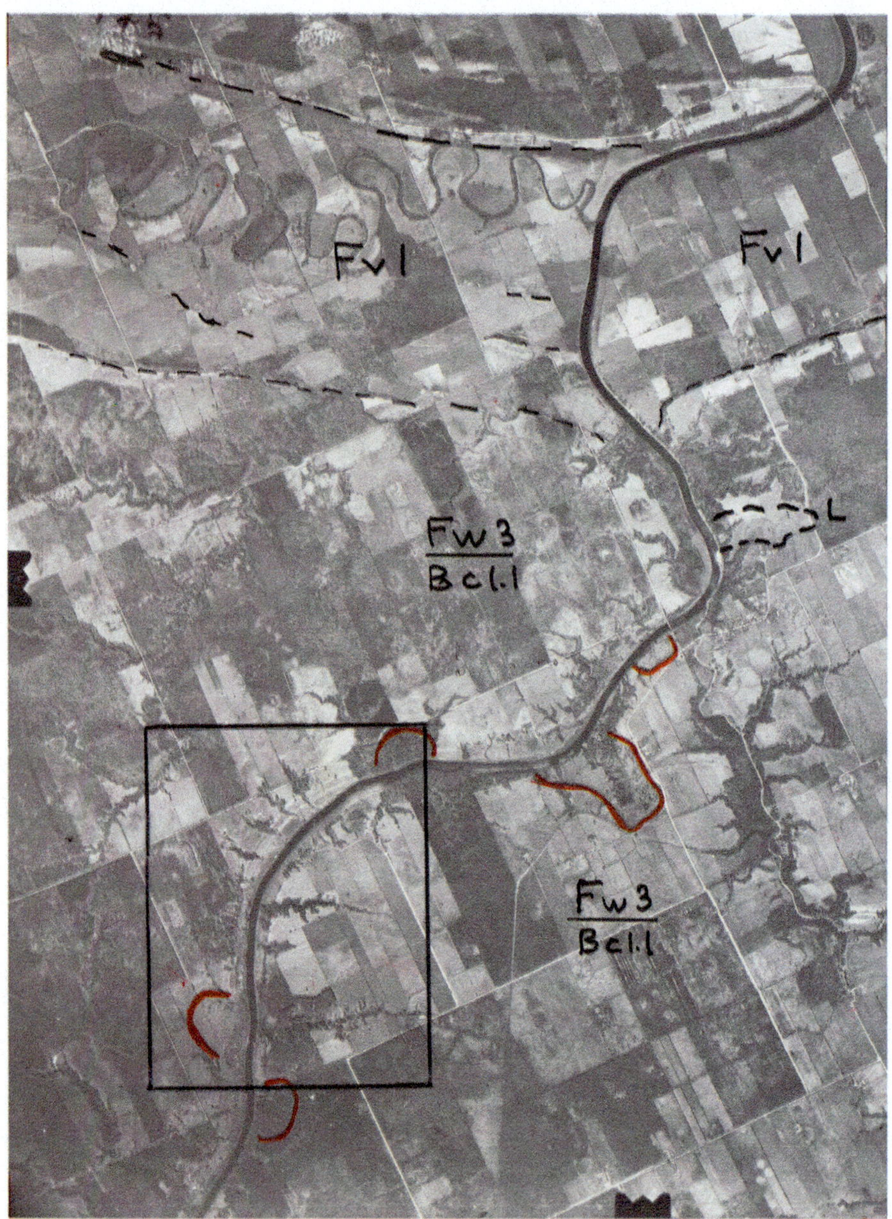

E19 Airphoto (B)

Photo Interpretation

This natural colour air photo is 1:10,000 scale.

The **Mf1**failure occurred on 16 May 1971 on a plane of **Bc1.1** soft clays 20 m below surface in a 43 m stratigraphic section of **Fw3** deltaic sand, 3.5 m of stiff clay, 21 m of soft clay, 4.5 m of stiff clay, 10 m of glacial till. Saturation of the pervious sands during spring snow melt permeated to the underlying clay, increasing its water content and pore pressures.

The natural colour photo taken three and half months later on 01 September shows 400 m of retrogression as a propagated shear zone of this type of sudden earth flow of ridges of blocks of intact clay remaining in the crater. Debris that flowed into the channel of the South Nation River blocked the flow and caused flooding upstream.

The accompanying B/W panchromatic photo at 1:40,000 scale was taken 7 years earlier, 28 May 1964. Five old flows detectable on the photo are delineated occurring along a 6 km reach downstream. The dashed outline **L** marks the site of the 680 m long crater of the Lemieux landslide that occurred on 10 June 1993 and impounded the river over 3 km. River incision into the spoil proceeded slowly for 4 months up to November. **Fv1** are fluvial sands and silts in a 10 m incised abandoned channel bordered by river-trimmed scarps. Meandering Bear Brook is an underfit stream.

Photo Source

Courtesy of National Air Photo Library, A18360, 213; A30362, 152

Figure E20: Plantagenet

Location 45°31′20″N, 74°57′11″W

Landsat Image T/M 4-5, 09 October 2011

Regional Environment

The image covers 2,430 km^2 of eastern Ontario and southern Quebec divided by the Ottawa River. The Quebec area north of the river is in the Grenville Province of the Canadian Shield. The entire Ontario area is in the western basin of the Champlain Sea which occupied the Ottawa and upper St. Lawrence lowlands of the St. Lawrence Rift System from 12 to 9.5 ka. During this period the basin was uplifted about 135 m through crustal rebound. The green and dark brown terrains are generally forested terrains of paleodeltaic sands. The cultivated areas are located on glaciomarine clays. The arrow locates the site of Figure E19, 17 km to the south

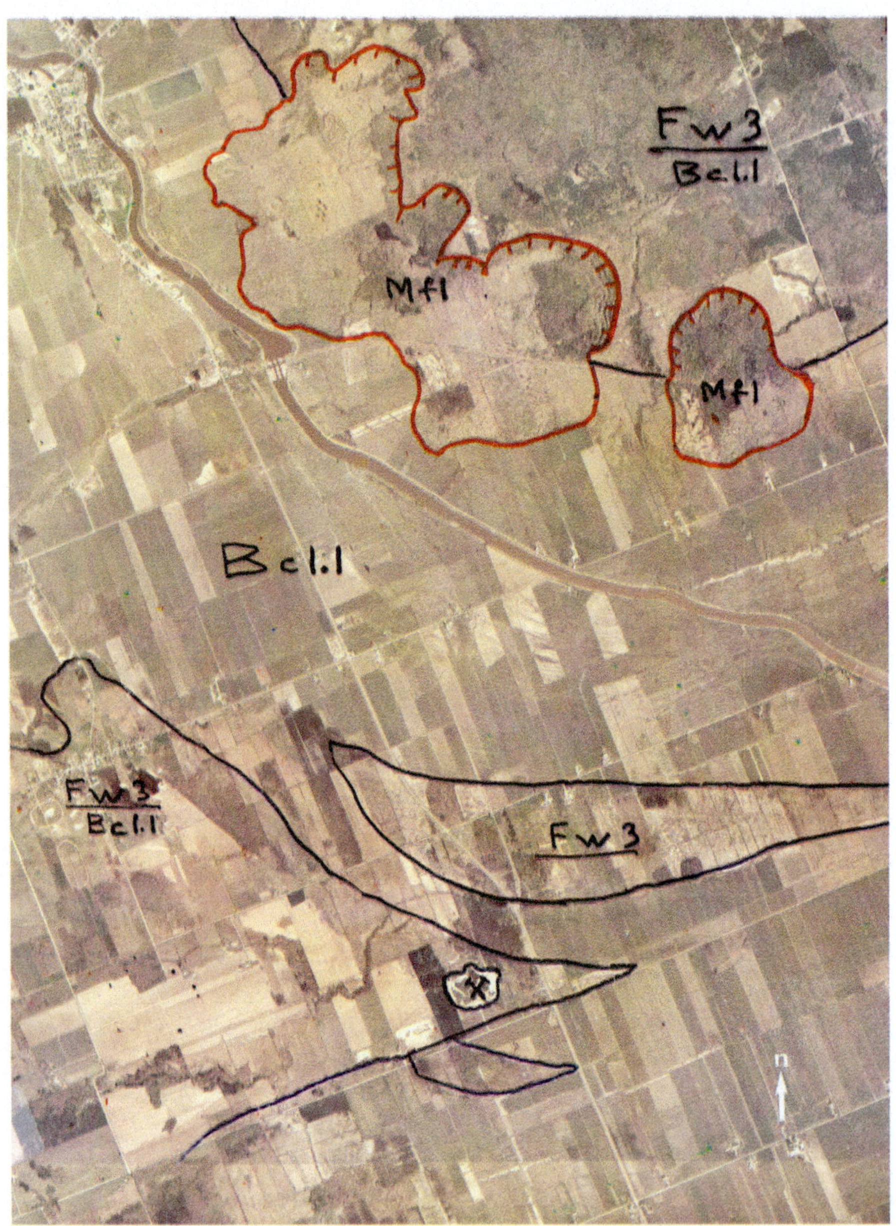

E20 Airphoto

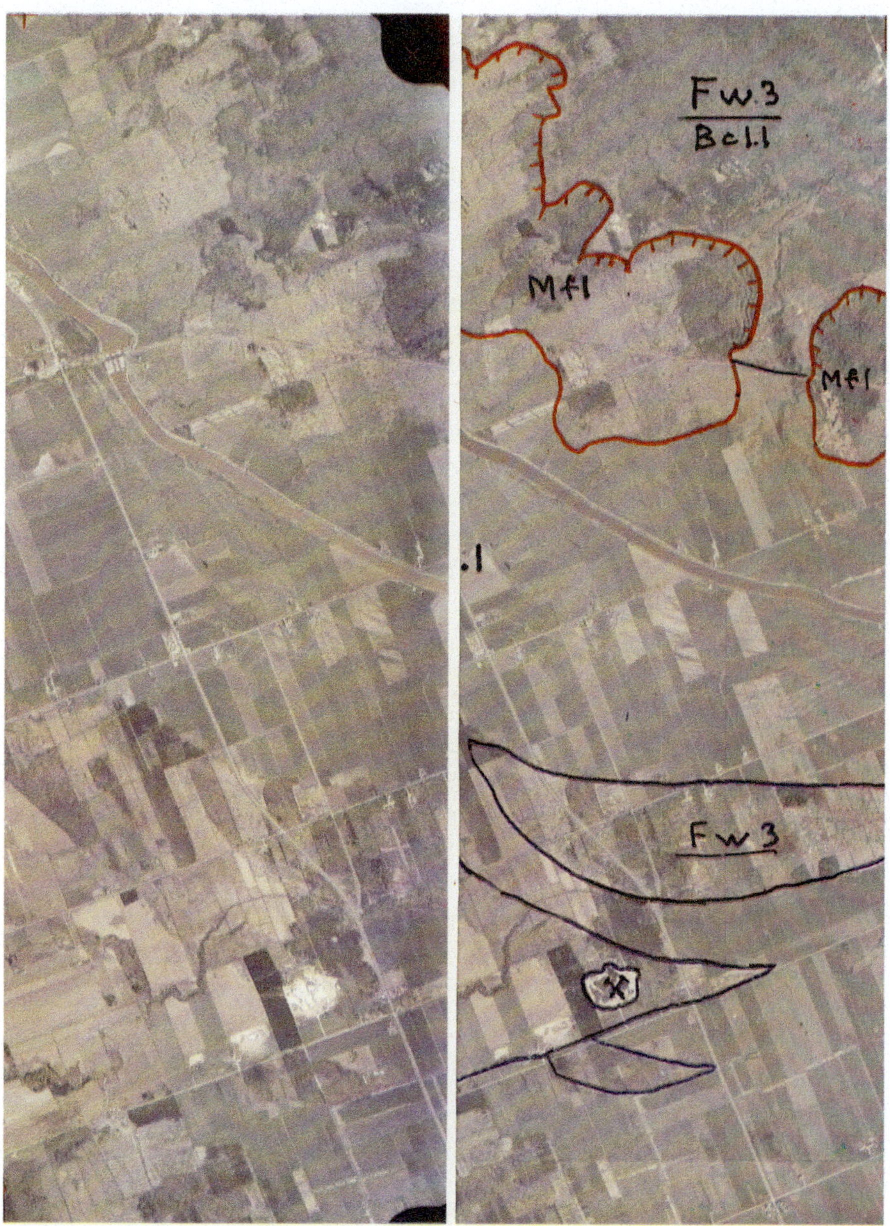

E20 Stereopair

Photo Interpretation

This natural colour air photo is 1:40,000 scale.

These multiple **Mf1** coalesced retrogressive earth flows are among a number of paleo-landslides in the **Bc1.1** Champlain Sea sediments that have been dated at *circa* 4,550 year BP.

As the scale of these earthflows is large and the relief subdued they are very difficult to distinguish on the ground. They were triggered by strong earthquakes during that period. Excess water migration from the permeable **Fw3** paleo deltaic sand layer to the impermeable clays also caused a strength reduction.

They are recognizable on air photos by the bowl-shaped scarps eroded into the sand terrace, and a characteristic "thumbprint whorl" pattern of block ridges on the surface of the spoil.

The flows retrogressed 600–800 m into the sand-capped terraces. With the exception of the northmost flow, the flow craters in the infertile sands are forested. The cultivated apron spreads out another 600 m from the failure bowl onto the clay sediments. A railway line crossing them is visible.

Flow marks of the ancestral river are faintly visible near the south bank of the present South Nation River. The town of Plantagenet is in the upper left corner of the photo.

Photo Source

Courtesy of National Air Photo Library, A31010, 61-62

The Comments and map of this Example and that of E20 are based on Geological Survey of Canada Open File 4475, *A review of the geology and geotechnical characteristics of Champlain Sea clays of the Ottawa River Valley with reference to slope failures.* J. S. Scott (2003).

Figure E21: Battle River

Location 52°23′54″N, 111°25′07″W

Landsat Image T/M 4–5, 08 October 2011

Regional Environment

The scene covers 2,475 km^2 in the Interior Plains of east central Alberta.

The Battle River arcs across a terrain of intensely cultivated hummocky moraine. The brown areas in the south are uncultivated arid grassland of flat shallow moraine draped over the shale bedrock.

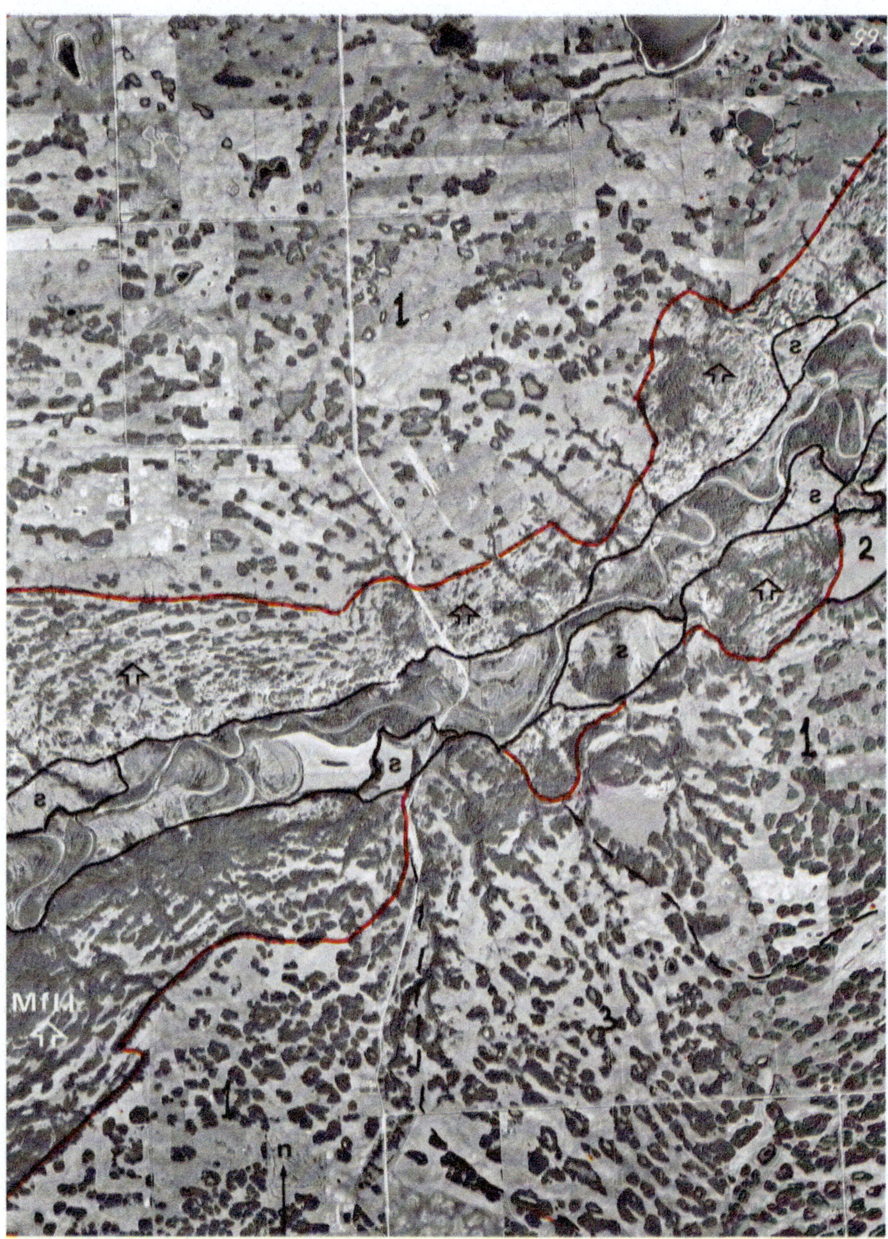

E21 Airphoto

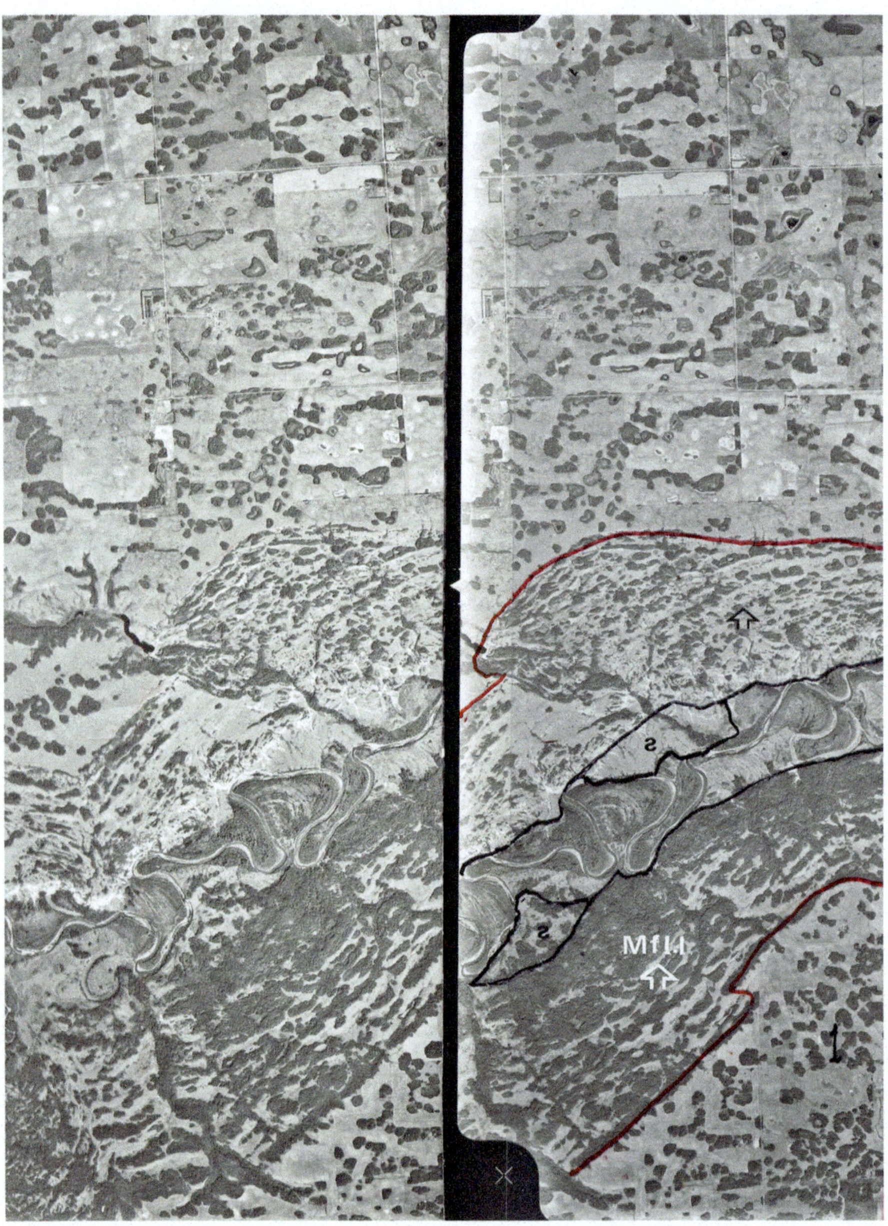

E21 Stereopair

Photo Interpretation

The meandering river in the 1:40,000 photo area is entrenched 55 m in glacial till and Cretaceous shale.

Deglaciation of the region occurred 12,000 years ago. Rapid postglacial stream erosion produced raised valley rims. Tension cracks formed in the thick glacial drift near the top of the valley walls. Progressive slips developed along the cracks in the flat-lying tills and the **Mf1.1** failed masses underwent a translational movement toward the river channel. Because all the flows indicated by **open arrows** along the river reach the bottom of the valley, the floodplain may also influence earlier stabilization of the toes.

Examination of satellite imagery shows the flows delineated on the photo extending for several kilometers along the sides of valleys underlain by Upper Cretaceous shales in Alberta and Saskatchewan.

Site elevations are 685 m at valley scarps, and 630 m at river channel.

S units are exposures of Upper Cretaceous shales in the floodplain.

Glacial Deposits
1—are **hummocky moraine**
2—is an area of **lodgment till**
3—is **ice-thrust moraine**

Comments are based on GSC Paper 66–37, Geological and engineering aspects of Upper Cretaceous shales in western Canada. J. S. Scott and E. W. Baker.

Photo Source

Courtesy of National Air Photo Library, A16490, 65-6.

Field Photos of Slope Failures

This selection contains 21 photographs taken by the author at 13 sites in the USA, Morocco, Austria, and Switzerland and one site map in Switzerland.

USA Sites

1: The Devil's Slide Complex Slide

The two photos are crossing the notorious slide which is located on the central California coast 25 km south of San Francisco. It is a large complex slide with which the state Department of Transportation has been battling for 75 years.

It consists of a series of many shallow slides, debris slides, flows, and rockfalls within larger slides, within an even larger slide.

The complex is approximately 760 m across and extends 275 m in height from below sea level to the top of the coastal cliff.

The slide is kept precipitous by wave undercutting on the beach below. The rocks are thick interbedded massive Paleocene marine sandstone, conglomerate, and siltstone that form the crest of the ridge, structurally overlying deeply weathered Mesozoic granodiorite.

L. Rivard and C. Hehner-Rivard, *Complex Terrain Mapping*,
DOI 10.1007/978-3-319-02450-9_3, © Springer International Publishing Switzerland 2014

The first photo shows the interbedded sedimentary rocks.

The second photo shows a roadway alignment at a point of local slope movement on the slide. Recent road closures have occurred in 1980, 1982, 1995, and 2006. Each of these closures lasted 4–5 months.

Since 2005 a tunnel of two single lane bores, 1,280 m long has been under construction to bypass this section of State Route 1. It was opened on 25 March 2013.

2: San Jose Rock Slump

The site of this photo is located on the east side of San Francisco Bay 15 km northeast of San Jose, California. The regional topography is controlled by a complex tectonic pattern. A Pliocene Orogeny uplifted and folded Miocene marine sedimentary rocks into a range, the Diablo Range, 450–750 m in elevation. A Mid-Pleistocene Orogeny accentuated the folds. Oversteepening caused by rapid uplifts resulted in landslides on a grand scale. In addition Diablo Range is bounded by two active faults. On the west, the Hayward Fault parallels the San Andreas strike 30 km westward, and on the east the Calaveras Fault. There is abundant evidence of mass movements due to earthquakes. The failure in the photo is a rock slump with well-developed block scarps at approximately 400 m elevation.

3: Point Reyes Debris Slide

Point Reyes is an island on the coast of northern California separated from the mainland by the continental San Andreas transform fault.

Geologically it is a composite section of moderately deformed Miocene marine shales and sandstones resting nonconformably upon a crystalline basement at altitudes averaging 60 m and rising to 250 m in the south.

The debris slide, with figures for scale, has a clear bulged toe. It is at approximately 70 m elevation.

Morocco Site

4: Ametrasse Rock Avalanche

This set of three photos, taken in August 1992, depicts a recent rock avalanche on the south flank of Jebel Akroud in the western Rif Atlas. The site is in Cretaceous flysch south of Tetuan Morocco.

The mass is 1.5 km long and 250 m wide.

This is an overall view of the avalanche mass.

This is a view upslope of subangular blocks from the distal limit of debris which crossed the roadway. The white house on the left is marginal to the mass and was affected on its foundation.

This is a close-up of block debris along the road. The house in background is beyond the failure mass.

Austrian Site

Creep Slopes

This group of four photos records the setting of seasonal creep on slopes of Mesozoic shales and slates of the Hohe Tauern in the village of Heiligenblut.

This photo shows the common direct evidence of creep by the deformed tree trunks on a forested slope.

This photo shows the typical angle of local slopes on a grass surface with stabilization staking in the background.

This is close-up of this type of biotechnical measure.

This shows the same measure on a gentler slope. Soil ruptures are visible on the margin of the staked zone and in the foreground of the scene.

Switzerland Sites

Grindelwald

This set includes nine photos taken on slopes above the village at elevations ranging from 1,200 to 1,400 m in the Alps of central Switzerland, 20 km south of Interlaken. The valley is noted for its high concentration of debris avalanches, slumps and slides. It was one of four alpine zones selected as test sites for the application of a special classification and mapping of geohazards; Kienholz, H. (1978). Maps of geomorphology and natural hazards of Grindelwald, Switzerland, scale 1: 10,000. *Arctic and Alpine Research, 10*(2), 169–184.

The local strata are Mid-Jurassic marly shales with some sandstone interbeds in the center of the Helvetic thrust and fold alpine nappes. Slides in these shales occur by saturation during snowmelt conditions.

As shown on the photo, these incompetent rocks form gently undulating slopes with little expression of internal structure. The region was covered to an altitude of 1,800 m during the maximum of the Würm glaciation 80–10 ka.

Debris Avalanche

This avalanche on an unconfined high slope is a matrix of unweathered blocks and finer weathered silt and clay size debris.

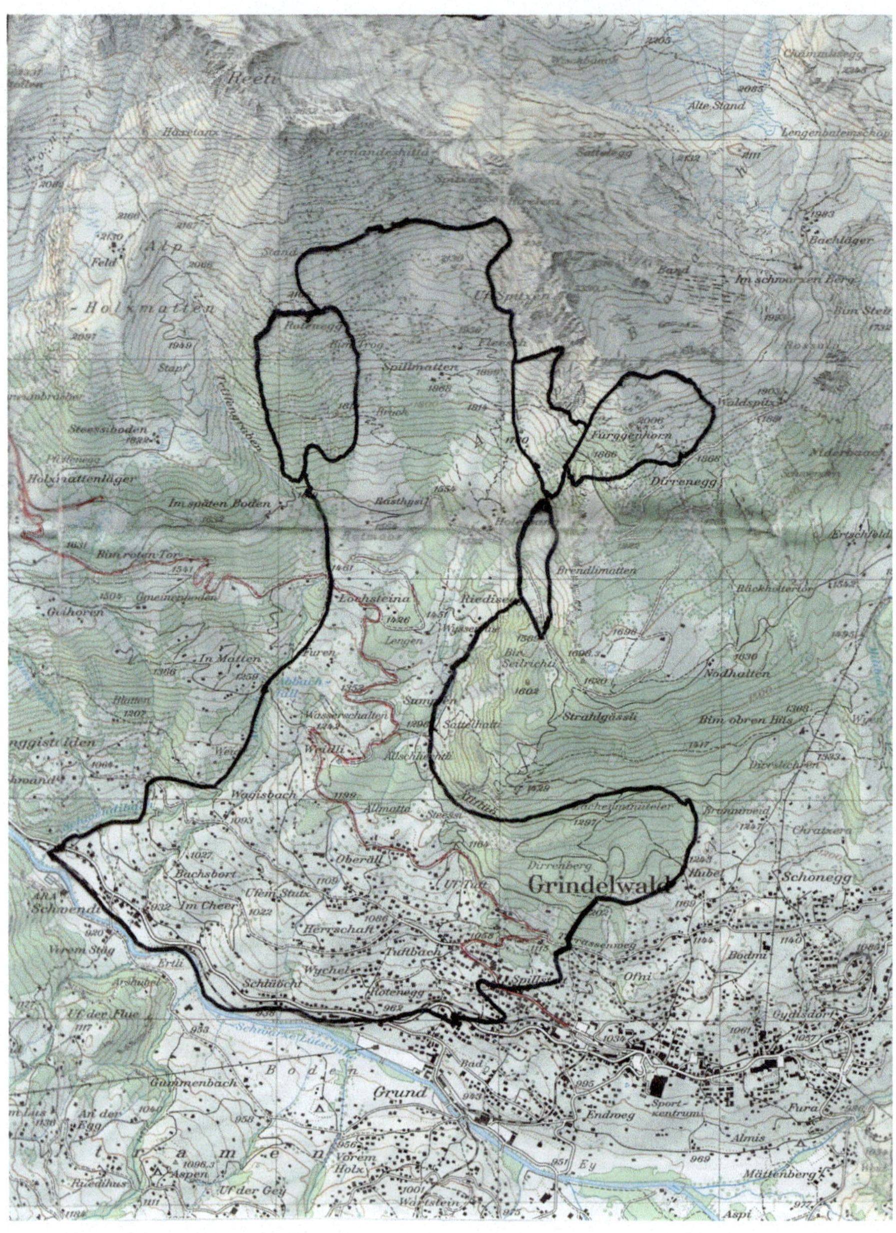

A 3 km long massive Holocene debris avalanche, partly urbanized and cultivated and partly forested, is delineated as mapped in 1938 on a portion of the 1: 25,000, 20 m contours topographic map of Grindelwald. The failure dropped 1,250 m from 2,200 m to 950 m at the river bank in the west end of the village. The unit is channelized 700 m wide upslope, spreading out to 2.5 on the unconfined lower slopes near the village. A 300 m zone of rockfall below a ridge scarp at 2,500 m lies above the head of the slide.

A 4 km section of an 8 km long access road to a mountain facility is marked in red. Its tortuous course reflects the hummocky relief of the debris mass that it traverses.

A group of three photos taken within the debris area show slabs of shale that had overtopped wire mesh cribbing stabilizing a roadway cut.

Debris Slides

This slide, composed of shale blocks, is on one of the higher slopes above Grindelwald.

Three slide occurrences are visible in this photo. The foreground unit is a composite slide. An upper movement (note two small trees) marked by its toe bulge is repeated below by a second with a large toe.

The third movement, beyond the small barn, appears more recent than the other two. It is notable by a larger bulge.

This is a general view of multiple slides whose relief is enhanced by some shadowing. The Obergletscher is visible across the valley in the background.

Debris Slump

A photo of a small failure with a slump block resting on its rupture surface

Geotechnical Failure

Construction activity above the retaining wall has provoked an outflow of shaly till from the subgrade of the main local mountain road.

The manufacturer's authorised representative in the EU is Springer Nature Customer Service Centre GmbH, Europaplatz 3, 69115 Heidelberg, Germany. If you have any concerns regarding our products, please contact ProductSafety@springernature.com

Printed and bound by CPI Group (UK) Ltd, Croydon, CR0 4YY
15/07/2026
02167624-0002